Between Bohemia and Sul

T0074721

This book identifies a distinctive kind of urban neighborhood that is on the rise throughout the USA, the dense, walkable, mixed-use bourgeois-bohemian suburb or the "boburb."

It looks at case studies of areas to live in Louisville, Kentucky. Based on scores of interviews with college graduates, backed by survey data and Census figures, it provides a clear, historical account of how these spaces arose. Chapters depict, analyze, and compare the Highlands neighborhood with other Louisville boburbs, contrasting them with the ephemeral bohemian quarters and the many suburban subdivisions. The Highlands are also compared with five other boburbs around the USA. Attention is given to the influence of transportation systems in shaping residential, community, and commercial spaces. Deeper cultural reasons for choosing the boburbs or the suburbs are also explored, including the political "big sort" between liberal and conservative places, and Bourdieu's account of how the distinction between economic and cultural capital shapes how people choose to live where they live.

This book will appeal to those interested in the evolution and distinctions among urban neighborhoods. It is ideal for academics and students within urban geography, urban gentrification, cities, and population.

William J. Weston is Van Winkle Professor of Sociology at Centre College in Danville, Kentucky.

Questioning Cities

The 'Questioning Cities' series brings together an unusual mix of urban scholars under the title. Rather than taking a broadly economic approach, planning approach or more socio-cultural approach, it aims to include titles from a multi-disciplinary field of those interested in critical urban analysis. The series thus includes authors who draw on contemporary social, urban and critical theory to explore different aspects of the city. It is not therefore a series made up of books which are largely case studies of different cities and predominantly descriptive. It seeks instead to extend current debates through, in most cases, excellent empirical work, and to develop sophisticated understandings of the city from a number of disciplines including geography, sociology, politics, planning, cultural studies, philosophy and literature. The series also aims to be thoroughly international where possible, to be innovative, to surprise, and to challenge received wisdom in urban studies. Overall it will encourage a multi-disciplinary and international dialogue always bearing in mind that simple description or empirical observation which is not located within a broader theoretical framework would not – for this series at least – be enough.

Urban Cosmopolitics
Agencements, Assemblies, Atmospheres
Edited by Anders Blok and Ignacio Farías

Urban Political Ecology in the Anthropo-obscene
Interruptions and Possibilities
Edited by Henrik Ernston and Erik Swyngedouw

Between Bohemia and Suburbia
Boburbia in the USA
William J. Weston

For more information about this series, please visit:
www.routledge.com/Questioning-Cities/book-series/SE0756

Between Bohemia and Suburbia

Boburbia in the USA

William J. Weston

Routledge
Taylor & Francis Group

LONDON AND NEW YORK

First published 2019 by Routledge

2 Park Square, Milton Park, Abingdon, Oxon, OX14 4RN
605 Third Avenue, New York, NY 10017

Routledge is an imprint of the Taylor & Francis Group, an informa business

First issued in paperback 2020

British Library Cataloguing in Publication Data
A catalogue record for this book is available from the British Library

Library of Congress Cataloging-in-Publication Data
A catalog record has been requested for this book

ISBN: 978-1-138-61456-7 (hbk)
ISBN: 978-0-367-73046-8 (pbk)

Typeset in Times New Roman
by Taylor & Francis Books

Contents

Illustrations

Figures

Tables

Acknowledgments

I am grateful to Centre College students Clara Gaddie and Ke Yao for their help gathering data. My Theory Camp members, Holly Couch, Stephanie Keller, Rachel Skaggs, and Meg Whelan, thoughtfully responded to an early draft. I got helpful feedback on an early draft from Elijah Anderson and his Urban Ethnography workshop at Yale, and from the Sociology Department of the University of Louisville. Phil Gorski and the Working Group on Human Flourishing, Social Solidarity, and Critical Realism helped shaped the material in "Are Neighbourhoods Real?" Nate Kratzer and Harrison Kirby, then with the Greater Louisville Project, were instrumental in handling the quantitative data on Louisville neighborhoods; I am particularly grateful to Harrison Kirby for making the maps. Qualtrics conducted the survey of the five boburbs outside of Louisville. The Center for Neighborhoods generously shared their knowledge and connections. Jack Trawick and Steve Wiser, two of the great experts on Louisville neighborhoods and architectural history, respectively, were especially generous with their knowledge of the city.

Beau Weston

Introduction

Walking to the coffeehouse

I teach a course about coffeehouses. I also teach this course *in* coffeehouses – and hold my office hours, and do my writing, and meet with people, and just hang out, in coffeehouses. When at home, and when I travel, I go every day to a local independent coffeehouse. They make my daily mocha, provide the buzz of human community, and locate me in an actual place in a real neighborhood.

I am not well versed in the nuances of coffee – I drink a mocha. I have no idea of the provenance or distinctive flavor of the espresso that is under the milk and chocolate. I am not a coffee snob.

I am a coffeehouse snob.

This kind of coffeehouse is a "third place" – not home, not work, but a place where strangers can become acquaintances (Oldenberg, 1989). A third place is open to all, and strangers are welcome. For the regulars, a third place fosters the kind of network of weak ties which holds a real community together (Granovetter, 1973). A good coffeehouse, and other hangouts like it, can turn a *space* into a distinctive and loved *place* (Warnick, 2016; Gieryn, 2000. For classic statements on place and space, see Tuan, 1976 and Relph, 1976). More than just a local network, third places are where the conversations can happen which create the public sphere, the face-to-face source of public opinion (Habermas, 1962).

I have noticed that independent coffeehouses, the kind that create a distinctive set of regulars and become community institutions, tend to flourish in a specific type of neighborhood. If you know a city, you can find the independent coffeehouses by going to the mixed-use, walkable neighborhood. It will probably be near other interesting and independent stores, the kind favored by the main culture producers. Or, to pursue practical sociology from the other direction, if you want to find the mixed-use, walkable, distinctive neighborhood in a city, look for the greatest concentration of independent coffeehouses.

Many people associate coffeehouses with *bohemia*, the kind of gritty, poor neighborhood that artists, musicians, writers, free spirits, non-conformists, transgressives, and just eccentric people colonize. The bohemian coffeehouse is a place of original art on the walls, live music, spoken poetry, and endless conversation. Bohemia as an imagined community has endured for almost two centuries. Bohemia as a real neighborhood, though, is ephemeral, floating around the cheaper and sketchier parts of the city as artists discover where they can afford to live, and then culture consumers follow and price them out. Bohemia is "always already over" (Lloyd, 2006, 237).

Bohemia as a kind of neighborhood is constructed in contrast to *suburbia*. I will have much more to say about suburbia hereafter. Here I just want to note one feature: the suburban coffeehouse, if it exists at all, is more likely to be a drive-to, a drive-through, and a chain store. The suburbs have few third places that welcome strangers to hang out and become acquaintances. Suburban socializing is more often by invitation and in a controlled space. The distinctive social space of the suburbs is the country club.

This is not the end of the story of today's coffeehouses, though. The Starbucks revolution has made specialty coffee and espresso-based drinks a common part of the diet of middle-class and upper-middle-class people throughout the world. The old bohemian coffeehouse had to be cheap enough for the starving artists to hang out in. Today's espresso vendors – who also offer chai, and smoothies, and the whole world of infused teas – are concentrated in a kind of neighborhood between bohemia and suburbia.

The independent coffeehouse neighborhood is what this book is about. The land of the "bourgeois bohemians" – the bobos – is what I will call *boburbia*.

Walk with me to Highland Coffee

Highland Coffee is a real coffeehouse in the heart of the Highlands neighborhood of Louisville, Kentucky. Within a half-mile radius of Highland Coffee is a rich mixture of stores, restaurants, arts venues, little houses, big houses, apartments, parks, schools, churches, and other things, on boulevards, cross streets, and alleys, under a canopy of old trees, all mixed together. We will take just six short walks to the coffeehouse. There are dozens of possible routes there, all within a half-mile radius. We will note the real businesses that line our route in 2019, though we know that small businesses come and go rapidly in densely populated, and intensely competitive, retail streets.

Highland Coffee is on the main commercial street of the Highlands, Bardstown Road. Let's start at the beginning of Bardstown Road, at the north end, closest to downtown Louisville and the Ohio River. For a few blocks, Bardstown Road and Baxter Avenue are one, before splitting at a triangle – Baxter on the west and Bardstown on the east. The point of the triangle is the heart of the Original Highlands.

The first block of the Original Highlands is dominated by bars, especially Irish bars. Molly Malone's Irish Pub, Flanagan's Ale House, and O'Shea's Irish Pub make the old history of this neighborhood clear. They compete with DiOrio's Pizza and Pub, Wick's Pizza Parlor and Pub, Baxter's Bar and Grill, the Outlook Inn, and El Taco Luchador, which describes itself as a "funky taqueria." Many of these pubs are live music venues, especially on weekends. Banners along the street proclaim it "restaurant row." Most buildings are two

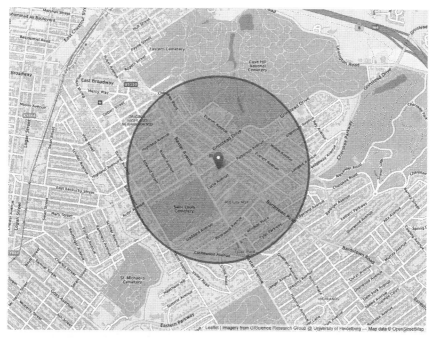

Figure 0.1 The dense environs of Highland Coffee

or three stories high, built a century or so ago, and have seen service as many, many different kinds of businesses over the years. At this moment in the life of Louisville, these few blocks are the heart of the bar-hopping precinct for college-educated young people.

Beyond the bars and restaurants, though, are many other kinds of businesses. There are three lunch places, in addition to the other restaurants and pubs. Many other services are provided – a bank, an insurance agent, an architect, "integrative veterinary services," lamp repair, four hair salons, a barbershop, a gym, a phone store, a therapeutic massage parlor, and two tattoo parlors. The large sign of a young woman with a colorful arm on Prophecy Ink Tattoo Studio faces across a small courtyard to a two-story mural on DiOrio's pub of a monkey in a chef's hat offering a large glass of the local Falls City Beer. You can buy fashionable women's clothing at Pitaya, tabletop games at Louisville Game Store, "fine gifts" at Celtic Jewelry, and all sorts of things at Eyedia consignment. In the middle of all this hubbub is Quills Coffeehouse. Except for the bank, the phone store, and the gym, none of these businesses are chain stores.

Even this intensely commercial block is not all for profit. The Episcopalian Church of the Advent serves its own religious community, and provides a food pantry and pet food pantry to the neighborhood. In the larger area the church participates in Highland Community Ministries. To serve the wider city, the church maintains a full-service St. George's Scholar Institute in the poor West End of Louisville, a couple of miles away. And to reach the wider world, the Church of the Advent works through the Highlands-based Kentucky Refugee Ministries.

Amidst and on top of all this commercial life are many apartments, and even some still-private houses. The block immediately behind these stores, across a small service alley, has other apartments, duplexes, and single-family houses stretching for blocks. Thousands of people live within easy walking distance of this lively, seven-day, long-hours commercial zone.

And all of this is just in the first three blocks of the Highlands. At that intersection, where Baxter and Bardstown split and Highland Avenue crosses both, there is a large mural proclaiming The Highlands, with the city's fleur-de-lis hanging above a road to a city on a hill. This is the first of several murals proclaiming Highlands pride – the most murals of any neighborhood in town.

Bardstown Road is lined with stores all the way to Highland Coffee and for miles beyond. In the very next block stands the black-and-white facade of Jack Fry's restaurant, an important landmark in the revival of the street from its scruffiest, most bohemian days in the 1970s. Restaurants are often the crucial pioneers in changing a neighborhood from a merely local place to a destination for outsiders. An old church houses Holy Grale pub and Gralehaus coffeehouse. It is across the street from Friend's Hookah Café, near the punning Yang Kee Noodle restaurant. A few doors down, next to a beauty academy, is Oneness Boutique, "*the* number one destination in Kentucky for all things premium in apparel and footwear ... a pseudo-mecca for all things culturally relevant within our area" (Oneness Boutique, 2019). Vintage Barber is brand new, but old style, next to the genuinely old Decades Antiques. The Somewhere restaurant shares space with the Nowhere Bar. There are many other eateries, tattoo parlors, smoke shops, and other services. There are also residues of Bardstown Road's industrial past: Halmar Corporation has manufactured boiler parts since 1950.

There are some ordinary chain stores in the next quarter mile – Wendy's, Taco Bell, Buffalo Wild Wings, Chipotle, and a Speedway gas station and convenience store. At the triangle with Baxter Avenue is a small chain mall – Walgreens, Great Clips, Highland Nails, Jimmy John's, and even a Starbucks – the only Starbucks in the Highlands. Recently some less-ordinary chains have also made a stir on Bardstown Road proper. Steel City Popsicles, Hop Cat craft beer pub, and Urban Outfitters clothing store have come to the Highlands to share in the neighborhood's distinctive vibe. This has led many longtime residents to fear that the Highlands is losing its soul, turning into another kind of upper-middle-class mall. The great majority of the stores on Bardstown and Baxter, though, are still local.

The Urban Outfitters brings us to Highland Coffee, as they occupy two ends of the same building. This one-story brick structure sits end-on to Bardstown Road, with a twenty-car parking lot serving both businesses. Before Urban Outfitters took the street end, there was a Blockbuster Video store there – a succession showing the fast-moving changes endemic to corporate capitalism. But the other end of the building also reveals the changes that come to "buy local" capitalism, too. Before the coffeehouse came, the space was occupied by a legendary record store, Ear X-tacy. When the expanding music store moved to a larger space further down Bardstown Road, Ear X-tacy's owner wanted to be sure that a suitable tenant was found for their old space. He embraced Highland Coffee as a worthy successor – which, indeed, it has proven to be for his customer base (Timmons, 2018). Ear X-tacy itself was forced to close in 2011 as the record business collapsed. It was succeeded by a Panera Bread store.

The "buy local" fans of Highland Coffee resent Urban Outfitters for a little insult, and for a bigger change of tone. The little thing was that the clothing chain made the local coffeehouse take down its permanent sign on Bardstown Road. Now, the coffeehouse must put a signboard out on the sidewalk every day, and take it in each night. The main emblem of Highland Coffee, a four-foot wide yellow coffee cup, still sits above the front entrance of the shop, but that door is well back from the street. They even started a sticker campaign, "I Know Where Highland Coffee Is," to fight back against this small act of marginalization. The larger resistance to Urban Outfitters, though, is part of the conflict between the actual bohemian types who live in the Highlands, and what they see as the routinization of the Bardstown Road vibe. Suburban mall kids, Highlanders believe, will come to the Highlands to soak up the thrift-store style, then buy overpriced knockoff versions from places like Urban Outfitters.

If you walk to Highland Coffee from the other direction, up Bardstown Road from a point a half-mile south, you would start at Day's Coffeehouse, which serves a slightly different neighborhood than Highland, as well as being something of a gay hangout. Indeed, the stretch of Bardstown Road from Day's Coffeehouse to Highland Coffee is home to the main gay-friendly bars in Louisville, notably Big Bar, Chills, and the above-mentioned Nowhere Bar (Lauer, 2017).

Walking north from Day's you will also pass a dense array of shops, services, a few community service institutions, and lots of places to eat. The Bristol Bar and Grille was an early agent of the Highlands' rise from bohemian seediness in the 1970s. Kizito Cookies and African Crafts is a notable local success story, created by a Ugandan immigrant woman whose cookies are sold all over Louisville. Another landmark is the two-story mural, on the side of the Wine Rack, proclaiming "The Highlands – Weird, Independent, and Proud."

At the intersection with Longest Avenue – named, not for its length, but for the early developers the Longest brothers – is an interlocked trio of independent businesses that symbolize the Highlands as well as anything could. On the corner is Carmichael's Book Store. In the back of Carmichael's, connected to it internally, was the first Heine Brothers coffeehouse, the dominant Louisville-only coffee chain. Next door is Ramsi's Café on the World, a fusion-everything restaurant. While waiting for a table at the restaurant, you are encouraged to take a buzzer and browse the bookstore until called. One reviewer painted this lovely picture of the cosmopolitan ideal here:

> Once Ramsi's finds a spot for you, you'll find yourself in a place where youngsters talk skateboards, hipsters talk piercings, erstwhile hippies talk portfolios, and nearly everybody talks politics – mostly in a progressive vein. [It] feels like it might have been decorated by a well-traveled scavenger … The menu … seems to have been scavenged from across the globe as well (Rosen, 2008).

The Heine Brothers store outgrew its place in the back of the bookstore, and built a spanking new shop across Bardstown Road. The Heine Brothers coffeehouse is now an outparcel of the Mid-City Mall. This is a mall, with a difference. Sure, it has a Subway, a nail salon, and a grocery store. But it also houses the local public library, a comedy club, and Baxter Avenue Cinemas, the main arthouse movie theater in Louisville.

This mile of Bardstown Road has the most interesting array of local businesses in Louisville. It is what makes the Highlands a destination for people from all over the city and the region. To "go Bardstown Roading" is a self-explanatory pastime, especially for young people.

However, the commercial side of the Highlands is only the half of the rich density of this most mixed of mixed-use neighborhoods. The residential neighborhoods on either side of Bardstown Road – Cherokee Triangle on the east, Tyler Park on the west – are themselves cherished *places* in the social sense. They each have a proud and active Neighborhood Association, with banners on the lampposts and placed in house windows. Unlike a suburban subdivision, they contain a rich array of kinds of residences. There are mansions and expensive condominiums on the big streets, elegant 1920s streetcar apartments, less elegant 1970s garden apartments, bungalows, shotgun houses, big houses cut up into tiny apartments, and garages and outbuildings retrofitted as "grandma" apartments. Nearly every residence is distinct from its neighbors, which adds to a greater whole characterized by mixture, variety, and the most common Highlands word, diversity.

The Cherokee Triangle is one of the most varied, interesting, and, today, desirable neighborhoods in greater Louisville. It is wedged between Bardstown Road on the west, and a pair of great green spaces on the east, Cave Hill Cemetery and Cherokee Park.

Let's start at the end of Highland Avenue, where it meets the cemetery wall. If we could see over the wall, we would be near the grave of the cemetery's most famous resident, Muhammad Ali. This is a leafy, quiet corner of the city – on both sides of the wall – even though we are a scant five blocks away from the jangle of the Baxter-Bardstown triangle. Walking toward Bardstown Road we pass a row of brick bungalows from the 1920s. Bungalows are small – one bedroom, one bath, with a low-ceilinged attic, about 1800 square feet. They were a mainstay of the streetcar neighborhoods like this one, and in those days would have held a whole family in modest comfort. Today they house smaller families, or probably just a couple or single person. They go for more than $200,000 (all current real estate data from Zillow.com).

At the next corner we see signs of children. The single-family brick house with the lovely detailing in the door glass, built in the early 1900s, has a soccer-ball themed planter on the wide porch. The house on the opposite corner, a big four-bedroom wooden house, built in the early days of the neighborhood in the 1870s, would sell for about half a million dollars. It has toy trucks in the yard. Right next door, though, we see how they achieve great density in what still feels like a single-family-home neighborhood. The brick single-family house has converted one floor to an income apartment. The three-story building next door has three apartments, but does not seem to loom large because the mature trees are even bigger than the building. Tucked away down the street was the end of a long, low, in-fill apartment building from the 1970s, with its hidden parking lot facing the cemetery wall.

Continuing down Highland Avenue toward Bardstown Road we come to one of the great do-good institutions of the city, Highland Presbyterian Church. We first see its wood-chipped and well-landscaped playground running along Highland Avenue. Across the street is its education and ministries building, which proclaims that it is the home to Kentucky Refugee Ministries, a major resettlement agency for the whole state. The church sanctuary and its attached classrooms, on the south side of the street, stretch from the service alley all the way to the next corner. The back half is a half-timbered addition, meant to suggest the rural English churches fashionable a century ago. The front of the church, built in 1887, faces the main boulevard of the neighborhood, Cherokee Road. The church is not the only service fitted in to this residential neighborhood – across Cherokee Road is the Carnegie library building, constructed 1899. It now houses a financial services firm with a discrete sign.

Directly under the permanent Highland Presbyterian Church sign is a temporary yard sign reading "No matter where you are from, we're glad you're our neighbor" in English, Spanish, and Arabic. This sign, with the text printed within distinctive green, blue, and orange stripes, is a common marker of welcome on houses, churches, and some businesses throughout the Highlands. It is often near cars with stickers for liberal causes and Democratic candidates. The cars are more probably foreign than domestic, and much more likely to be sedans or minivans than SUVs or pickup trucks. Hybrid cars are common. Conservative signs of any kind are rare.

Heading south on Cherokee Road, running parallel to Bardstown Road on our way to Highland Coffee, is one of the prettiest streets in Louisville. The sidewalks are wide, and the grassy skirt between sidewalk and street is wider still, filled with large, century-old trees. This was the street of big houses of the Gilded Age. Over the decades it also acquired beautiful three-to-five story walk-up apartment buildings from the early 20th century, more modest single-family homes, and some modern apartment buildings. It was when developers started tearing down the mansions and building long, skinny, apartment buildings in the 1960s and early 1970s that the residents mobilized to save the Cherokee Triangle.

At the next intersection, where Grinstead Avenue crosses Cherokee Road, Highland Baptist Church, built 1893, and its related education buildings, from the 1950s, fill one corner. Begun as a mission of Southern Baptist Theological Seminary, located a few miles away, Highland parted company with the increasingly conservative Southern Baptist Convention in the 1980s. Like Highland Presbyterian, it sports a three-color "we're glad you're our neighbor" sign in front of the church.

The next block brings us to Patterson Street. Here we find another mainline Protestant church, Highland United Methodist. While it does not have a three-color yard sign out front, it does have a permanent banner attached to the building proclaiming the diversity theme common to the neighborhood: "Building a diverse family together in faith and love." As we turn right (west) on Patterson, we can see Bardstown Road, two blocks away. When we get one block closer, we can see in the distance the dome atop the tallest building in Kentucky, the 35-story 400 West Market building, some three miles to the north in the central business district.

And one block further west brings us to Highland Coffee.

If we started a half a mile southeast of Highland for our walk in, on Willow Street, we would see several multistory condominium buildings overlooking one entrance to Cherokee Park. Though these buildings are some twenty stories tall, the hilliness of this part of the Highlands means they do not dominate the streets of houses around them. A large-three-bedroom condo, overlooking the park, goes for well over a million dollars.

Willow Park is a tiny spear-tip of the larger Cherokee Park, reaching into the center of this section of the Cherokee Triangle. It has a small enclosed playground and a grassy area that is home to live music in the summer. This little park is the heart of a micro-place for young families, a walkable respite that is not just an amenity, but a life-saver for the stroller set.

The long streets that go between the park and Cave Hill Cemetery have imposing mansions. Willow Street, a smaller cross street, runs uphill from the little park. It has a more mixed set of lovely single family houses from the 1890s, with big porches, interesting detail in the wood work and in the door glass, and lots of care in the paint and landscaping. This was built as a neighborhood for established families. The Victorian houses have three or more bedrooms on two or three floors, set close to neighbors, and close enough to the street to talk from the porch to the passersby. At the end of the street are two- and three-unit apartment buildings from the 1910s, and a larger complex of two-story apartment buildings from the 1970s. These are adjacent to the single-family homes, in the same leafy, quiet neighborhoods. Yet, when we turn left (west) on Patterson Avenue, we are only three long blocks from our destination, Highland Coffee.

If we walked to Highland Coffee from the southwest side, in the Tyler Park neighborhood, we could start from Tyler Park itself. The park is a triangle of green with tennis courts, a playground, and a pretty stone arch carrying the park under busy Baxter Avenue. The houses overlooking the park are wooden three-story family houses, in various styles, all built around 1900. The blocks in between the park and the coffeehouse are packed with similar century-old houses, some later infill, and a few small apartment buildings. Near the street leading to the coffeehouse is St. Louis Cemetery, a 19th century Catholic burying ground. The far side of the cemetery, another block west, is Barrett Avenue, which forms the boundary between the Highlands and the more working-class Germantown.

Finally, if our walk began at the northwest edge of our half-mile circle, at the intersection of Barrett and Winter Streets, we would be in a scruffier commercial area of bars, liquor stores, and inexpensive restaurants. The houses on Barrett are smaller and less renovated than those on the Cherokee Triangle side of the Highlands. The closer we get to Bardstown Road, block by block, the more artistically painted the houses get. The liberal political stickers on cars multiply. Many yards have the three-language "We're glad you're here" signs, as well as "Black Lives Matter" and, in election seasons, liberal candidate placards. Several houses have small personal book-lending libraries on poles by the sidewalk.

From either the southwest or the northwest, eventually we reach Patterson Avenue. On this side of Bardstown Road, Patterson Avenue is not the leafy, broad street that we enjoyed on the Cherokee Triangle side. It is more like an alley, serving the parking spots and garage apartments of the houses on the parallel streets. The main occupant of this block of Patterson Avenue is Bloom Elementary School. The school sits a literal stone's throw from Highland Coffee, and there is a tight connection between the two. Bloom teachers often meet in the coffeehouse, and art from the children adorns the walls of Highland Coffee. Indeed, some of the Highland Coffee regulars were active in the citizens' movement to save Bloom Elementary when it was threatened with closure. Bloom has become a treasured public institution to many Highlands parents.

These six walks – down and up Bardstown Road, from the northeast and southeast through the Cherokee Triangle, or from the southwest and northwest through Tyler Park – are just a few paths to Highland Coffee. The richness and variety of the neighborhood would be hard to exhaust, especially as it changes all the time in the vibrant ecology of a vital part of the city. These walks are short – just half a mile, each one.

Moreover, we could repeat this experience of the rich, mixed-use, vibrant life of the Highlands in several more, non-overlapping, mile-wide circles as we headed southward down Bardstown Road until we got to today's suburbs. It would be hard to find any other place in greater Louisville with such a jam-packed, diverse, walkable neighborhood. Of all the crucial features of the Highlands, this dense richness is the most distinctive.

References

Thomas Gieryn, 2000. "A Place for Space in Sociology." *Annual Review of Sociology*, 26: 463–496.

Mark S. Granovetter, 1973. "The Strength of Weak Ties." *American Journal of Sociology*, 78, 6: 1360–1380.

Jürgen Habermas, 1962. *The Structural Transformation of the Public Sphere: An Inquiry into a Category of Bourgeois Society*. Translated by Thomas Burger. Cambridge: MIT Press, 1989.

Landon Lauer, 2017. "An Analysis of Gentrification's Effects on LGBTQ+ Populations in Louisville, Kentucky." University of Louisville, College of Arts & Sciences Senior Honors Theses. Paper 128.

Richard Lloyd, 2006. *Neo-Bohemia: Art and Commerce in the Post-Industrial City*. New York: Taylor & Francis.

Ray Oldenberg, 1989. *The Great Good Place: Cafés, Coffee Shops, Bookstores, Bars, Hair Salons and other Hangouts at the Heart of a Community*. New York: Marlowe and Company.

Oneness Boutique website, 2019. https://www.oneness287.com/pages/about-us.

Edward Relph, 1976. *Place and Placelessness*. London: Pion.

Marty Rosen, 2008. Restaurant review of Ramsi's, *Courier-Journal*, May 3, S18.

John Timmons, 2018. Interviewed in a Centre College student film.

Y.-F. Tuan, 1976. *Space and Place*. Minneapolis: University of Minnesota Press.

Melody Warnick, 2016. *This is Where You Belong: The Art and Science of Loving the Place You Live*. New York: Viking.

Part I

The Louisville Highlands

An exemplary boburb

1 Louisville

The big picture

The core of this book is an empirical study of Louisville, Kentucky. Today in Louisville the Highlands neighborhood is an exemplary boburb. We will see how the Highlands developed into a boburb at a specific moment in the history of the city and of the neighborhood. But we will also see that a city is an ecology, in the manner Jane Jacobs described, always changing and developing (Jacobs, 1961). The Highlands was not always a boburb, and probably will not always be one. We can see this by looking at the neighborhood's past, and by examining the potentially subversive trends already at work in the Highlands today.

We can also see what a boburb is by comparing the Highlands with the neighborhoods most like it in Louisville today. Other neighborhoods are defined more by their contrast with the Highlands. Crescent Hill is also a boburb, though perhaps more bourgeois than the Highlands. Germantown and Clifton are new bohemias because they are cheap enough and urban enough to draw young artists. Barbourmeade is a classic post-World War II automobile suburb. Lake Forest, a 21st century gated community, is in many ways the anti-Highlands. Norton Commons is a complete "new urbanist" community, modeled on the Highlands, built at the far edge of the county. In NuLu and Portland, a concerted effort is being made to create instant bohemias, hip neighborhoods based on creative enterprises. Finally, Russell and Shawnee are currently an African-American ghetto, but may have the bones to be a multicultural boburb of the future (for an ecological account of Louisville neighborhood development, see Wiser, 2008).

I focused this study on college graduates, because they have more choice about where to live. There are few very poor people in this study – though some of the artists choose to live on very little. There are few very rich people in this study – though some are clearly at the high end of the upper-middle class. The heart of this account comes from more than 180 qualitative interviews of leaders, experts, activists, and ordinary residents. The insights from these interviews are supplemented by my own survey of some 500 Centre College alumni in Louisville; from Census data; and from other quantitative databases. I also gathered survey data from 250 residents of boburbs in Austin, Chicago, Philadelphia, Atlanta, and Portland for comparison.

Louisville is a city of 750,000, the center of a metropolitan region twice that populous. The places to know for this study are:

- *Downtown* – the business district, centered on the Ohio River;
- *NuLu* – a commercial area just east of Downtown, now being turned into a creative center;
- *Portland* – a working-class neighborhood just west of Downtown, which some people are laboring to turn into a creative center;
- *Russell and Shawnee* – a ghetto just south of Portland in a formerly middle-class streetcar suburb;
- *Germantown and Clifton* – working-class neighborhoods just outside downtown, now becoming bohemian;
- **The Highlands** – a mixed-use boburb just beyond Germantown, surrounding Bardstown Road;
- *Crescent Hill* – a mixed-use boburb heading east along Frankfort Avenue;
- *The Watterson* – Interstate 264, a perimeter highway, often taken as the divide between city and suburb;
- *Barbourmeade* – a car suburb outside the Watterson;
- *The Snyder* – Interstate 265, a further-out perimeter highway, taken as the divide between suburb and the still-rural parts of the county;
- *Lake Forest* – a gated community outside the Snyder, near the eastern edge of the county;
- *Norton Commons* – a new urbanist community at the eastern edge of the county.

Louisville began on the south bank of the Ohio River, and initially developed westward along its course. Over time, though, the direction of development shifted eastward. One reason was that the western movement ran up against a sharp southward bend in the Ohio River. At the same time, Kentucky's "golden triangle" among Louisville, Lexington, and the Cincinnati suburbs pulled the development of the richer residential neighborhoods to the east. The railroad line headed east to Frankfort, the state capital. Frankfort Avenue, running along the train line, was an early streetcar line. Lexington Road, one of the many names of state route 60, was the main east–west road in Kentucky before the creation of Interstate 64 after World War II (Figure 1.1). Thus, the residences of the richer people in Louisville have continued to shift east as the streetcar lines were displaced by the automobile, and the railroad line was displaced by the interstate highway (Figure 1.2).

The second factor that interfered with the even spread of Louisville was racial segregation. As Figure 1.3 shows, the neighborhoods immediately southwest of downtown are heavily black. This partly reflects the early pattern of segregation. Russell, the "Harlem of Louisville," was separated from the white grandees of Old Louisville. The black working class of Parkland and California (the "far west" of the city) were divided from the white working class of the ethnically marked Germantown and Limerick. Still, at

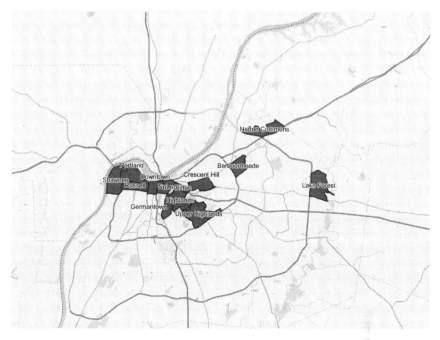

Figure 1.1 Louisville neighborhoods and interstate highways

Legend
- 0
- 15
- 30
- 45
- 60

Figure 1.2 Percent low income (150% of poverty line) in Jefferson County, by 2010 Census tract

Figure 1.3 Percent African American, 2010 Census

the beginning of the 20th century, there were white middle-class neighbor-hoods along the streetcar line to Shawnee Park in the west end. After the race riots of the 1960s and the white flight of the 1970s, though, the west end became the classic black ghetto of Louisville. The "dirty south," always more working class than the east end, has become the main immigrants region of the city.

Note, though, that the black tracts and the poor tracts are not the same. Right in the middle of Jefferson County, outside the limits of the old city of Louisville, is Newburg. This was based on the free black community of Peterburg even before the Civil War, and it remains a node of black bourgeois culture today. As Figure 1.2 shows, black and immigrant Newburg and Bue-chel are poorer than the surrounding tracts, but they have a working-class level of average poverty – not an underclass ghetto.

Education is increasingly the driver of socioeconomic status. Income is as much an effect of education as it is a cause. The great achievements of the civil rights movement over the past two generations means that race is declining in significance in determining the life chances of African Americans (Wilson, 1978). Today, the bachelor's degree is the great dividing line in the social class structure of the United States. The education map (Figure 1.4) shares a rough connection to the income map (Figure 1.2), and an even rougher connection to the race map (Figure 1.3). *The concentration of edu-cated people in the boburbs is the core feature of their cultural meaning.*

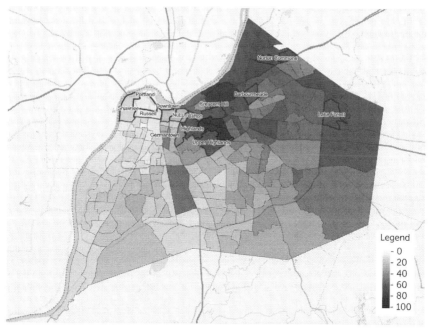

Figure 1.4 Percent Bachelors degrees, 2010 Census

College graduates mostly settle in a band from the Highlands eastward. While the whole metro region has a 28% college graduate rate, these tracts are more than twice that rate. Moreover, the far eastern suburbs are sparsely settled – rich subdivisions of college graduates were built amidst farms. The Highlands, by contrast, are a high concentration of college graduates right next to other dense census tracts with few college graduates.

The boburbs are not cheap. While the median house cost in all of Louisville is about $150,000, in Crescent Hill it is $231,000, while in the Highlands it is $281,000. This is in line with fully suburban Barbourmeade at $240,000, though considerably less than Lake Forest, where the median house costs $360,000. The Highlands has a greater variety of housing types, including many more apartments, than do the suburbs, producing a wider class range around the median. The choice among the several kinds of communities turns on something other than sheer cost.

An organization of young professionals gives us a finer-grained look at where socially engaged young college graduates in business and the professions settle (confidential personal communication; Table 1.1). An analysis of the zip codes of more than 750 members shows the great bulk live in a path from downtown eastward to the edge of Jefferson County, spilling over into adjacent Oldham County. Zip codes do not distinguish between boburb and bohemia, but they do let us compare the walkable neighborhoods with the car suburbs further out. Right now, fully a third of the young professionals live in

Table 1.1 Residences of young professionals in greater Louisville

Neighborhood type	% of young professionals
Downtown	9
Walkable (boburbia + bohemia)	22
Eastern post-war car suburbs (to the Snyder)	23
Eastern outer suburbs (including Oldham County)	13
Southeastern Jefferson County	11
Southern Jefferson County	6
West End	2
Other adjacent Kentucky counties	2
All other Kentucky residences	2
Southern Indiana	8

the dense, mixed-use part of Louisville. It is likely that as the young professionals become more established in their careers, and a higher proportion marry and have children, a significant proportion of the group will shift to the suburbs. Some, though, will stay to raise their children in the boburbs.

Political party registration gives us a rough measure of the ideological distribution of a neighborhood (Table 1.2). Bohemia and boburbia have almost three times as many Democrats registered as Republicans. Crescent Hill shares precincts with the more suburban St. Matthews, just inside the Watterson highway, which shades into the evenly divided suburban neighborhood, Barbourmeade, just outside the Watterson. In the far reaches of the county, beyond the Snyder, Lake Forest sits in the most Republican census tract in Jefferson County. Note, though, that none of these neighborhoods are totally one-sided in their partisanship.

In the 2016 election, Democrat Hillary Clinton won all of the bohemian and boburban neighborhoods. In one precinct of Germantown she received 91% of the vote. Clinton and Republican Donald Trump split suburban Barbourmeade. The far eastern gated community of Lake Forest, on the other hand, went strongly for Trump, giving him more than two-thirds of their

Table 1.2 Political party registration by neighborhood

	% Democrats	% Republicans
Germantown	67	22
Highlands	66	23
Clifton	65	25
Crescent Hill	55	36
Barbourmeade	47	43
Lake Forest	30	63

votes. The far-left Green Party candidate, Jill Stein, got her highest vote total anywhere in Jefferson County from a Highlands precinct.

The social density of the boburb rests on its greater physical density. In the Highlands and Crescent Hill there are about 6,000 and 5,000 people per square mile, respectively. By contrast, Barbourmeade, a 1960s car suburb outside the Watterson, has about 2,700 people per square mile. Out beyond the Snyder, in the gated community of Lake Forest, there are only 1,500 people per square mile.

The rise, fall, and rebirth of the Highlands

Over the two-and-a-half centuries of Louisville's history, the city has spread into the Kentucky countryside from its base along the river. About every half century it has grown to a new perimeter road, and then gone beyond it. The Highlands were born in the city's second 50 years as an outer suburb, beyond Broadway. The original Highlands grew fitfully in the second 50 years, capped by Eastern Parkway. A boom followed in the third period, when the Highlands name was a popular brand claimed by the new outer suburbs all the way to the Watterson. The fourth half-century saw a decline in Highlands as it was transformed into an inner suburb, while the new growth went eastward to the Snyder. In Louisville's fifth and current half century, the Highlands have been reborn as a fashionable and famous boburb.

The oldest parts of the Highlands residential neighborhoods were built by upper-middle-class families on the boulevards, and middle-class families on the cross streets. The families tended to be White Anglo-Saxon Protestants (WASPs), most of them recently from country towns. The most significant non-WASP groups in Victorian Louisville were the Irish and the Germans. The Irish were Catholics; the Germans were divided among Catholics, Protestants, and Jews. Some of the most economically successful Irish and German families also bought or built in the Highlands – especially if they were Protestant. Some African Americans did live there, as well, usually in the alleys. They tended to work as servants in white households in the neighborhood.

In the early 20th century, Louisville became more racially segregated, with most African Americans pushed into the west end. Segregation was required in all schools, and "petty apartheid" spread to other aspects of life. In mid-century, federal maps, estimating the credit-worthiness of the different neighborhoods in each American city, drew a red line around black neighborhoods. "Red-lining" by banks and real estate agents then dried up investment in those neighborhoods, and steered white home-buyers and business owners away from them (Poe, 2015). In the 1960s the Civil Rights movement finally put an end to legal housing discrimination. At the same time, though, race riots drove most white people and middle-class black people out of the black neighborhoods. Urban renewal and interstate highways crushed many poor neighborhoods, replacing them with projects that grew worse and worse, while big roads walled them off from richer places.

As many Germans and Irish became middle class in the 20th century, they moved out of Germantown, Limerick, and other ethnic enclaves close to downtown, into the then-suburbs, including the Highlands. By the 1950s the Highlands had become the center of Jewish life. The returning white GIs filled the new automobile suburbs with Baby Boom families.[1] These new subdivisions had covenants that excluded black people. The racially inequitable administration of GI Bill benefits also kept black GIs out of the new suburbs. The Highlands were somewhat more welcoming to all kinds of newcomers, including African Americans. This was due, in part, to the fact these neighborhoods had been built before "deed restricted communities" became the suburban norm, and in part because the people who stayed in the city tended to be more liberal.

In the 21st century, the ideological "big sort" became more pronounced (Bishop, 2008). Liberal people tended to be drawn to more urban neighborhoods that made a public virtue of being tolerant and diverse. Conservative people tended to move to the further and further-out suburbs and exurbs, as far from the city as their occupations allowed. The Highlands became known as the most liberal part of Louisville, the bluest spot in a red state.

In the 1960s and 1970s the Highlands did decline for a time. This was its moment as the bohemian quarter of Louisville. Young people seeking every kind of artistic life flocked to the apartments and divided-up old houses. Venues for live music, theater, art shows, comedy, and sheer cultural camaraderie opened up in bars, restaurants, and other commercial buildings on Bardstown Road. The Highlands became the hip and happening place. The success of the Highlands as a cultural scene then drew people with more money who wanted to enjoy the culture, while keeping their day jobs. These are the people we would later call the bobos, the bourgeois bohemians.

The decline of the Highlands drew another kind of urban type, the developer of cheap apartments. It was the threat of tearing down the old houses or dividing them up for transients that finally mobilized the residents of the Highlands to organize and protect their community. In doing so, they created a new consciousness of the Highlands as a distinctive kind of neighborhood.

I believe it was this process – of displacing the bohemians and developing a distinctive neighborhood consciousness – that turned the Highlands into a boburb.

Note

1   GI, probably derived from "Government Issue," is a common nickname for U.S. soldiers, especially of the World War II generation. After the war, the federal law known at the "GI Bill of Rights" subsidized college education, mortgages, and small business loans for returning military service members – skewed toward the white men.

References

Bill Bishop, 2008. *The Big Sort: Why The Clustering of Like-Minded America Is Tearing Us Apart*. New York: Houghton Mifflin Harcourt.

Jane Jacobs, 1961. *The Death and Life of Great American Cities*. New York: Vintage Books (Random House).

Joshua Poe, 2015. "Redlining Louisville: The History of Race, Class, and Real Estate."
   https://lojic.maps.arcgis.com/apps/MapSeries/index.html?appid=e4d29907953c4094a
   17cb9ea8f8f89de
William Julius Wilson, 1978. *The Declining Significance of Race: Blacks and Changing
   American Institutions.* Chicago: University of Chicago Press.
Steven Wiser, 2008. *Louisville 2035: A Look at What Louisville Might Be in 25 Years.*
   Louisville: AIA Publications.

2 Boburbia

The vibrant mixed-use community

The social type most at home in independent coffeehouses are "bourgeois bohemians" – bobos (Brooks, 2000). They have a bourgeois work ethic, but a bohemian cultural taste for the diverse and unusual. As an ideal type, bobos are adults who are defined by their work and leisure. But where do they live? What do bobo families value? What is a bourgeois bohemian neighborhood like? This study is an exploration of the *boburb*. The boburb offers walkable neighborhoods, mixing dense blocks of houses with small independent businesses, along with schools, libraries, and religious congregations. Boburbs are found in most American cities, often developing in the former streetcar suburbs. The boburb is a happy medium between the funkiness and grit of bohemia and the controlled safety of suburbia.

A national survey frames the choice well:

> Imagine for a moment that you are moving to another community. Would you prefer to live in …
> 1. A community where the houses are larger and farther apart, but schools, stores, and restaurants are several miles away [OR]
> 2. A community where the houses are smaller and closer to each other, but schools, stores, and restaurants are within walking distance?

This question helps distinguish those who prefer the suburbs or the boburbs. Given those two choices, Americans are *evenly split* – 49% choosing the suburbs, 48% choosing the boburbs (Pew Political Polarization Survey, 2014).

This survey reveals another important fact: most people can readily answer this question. Nearly everyone can envision these alternatives, and knows which side they are on. I asked everyone I interviewed which they would prefer, and none had trouble choosing. Indeed, for most people the choice was an immediate gut reaction. Some shivered when contemplating living in the other place.

In many cities, the old streetcar suburbs have become the ideal "boburbs." While the streetcars in many cities are long gone, the neighborhoods developed to serve them endure. They have several features which help make a rich and flourishing social experience for the people who live there, including children. The "traction magnates" who made their fortunes from streetcars often made most of their money from subdividing and selling the lots around the streetcar lines. These neighborhoods did not simply grow organically from the existing city, but were carefully planned as a whole. In particular, the streetcar suburbs were designed to pack enough houses in walking distance of the streetcar line to provide a viable ridership; to have businesses along the route that riders could shop in while walking to and from the streetcar; and to extend the city by connecting with the existing grid. Together these features contribute to a happy *social density* for a walkable residential neighborhood (on the value of social density, see Durkheim, 1893).

The streetcar neighborhoods have a mix of housing sizes and types. While the richest commuters could afford to live further out in the railroad suburbs, the professionals, middle managers, and secure working-class men and (some) women rode the trolley to work. Some of the houses, especially on the long streets parallel to the tracks, were bigger and made of more durable materials. Others, especially on the shorter crossing streets, were likely to be wooden "balloon frame" houses, built from standard designs and even pre-cut kits, selected by each homeowner. Later, more rationalized streetcar developments settled on some standard house plans, often bungalows. These houses were separated from their neighbors, but built close together. They normally had porches. They were set back from the street enough to have a little lawn – which became a great symbol of the American middle class – but close enough to the sidewalk that a pedestrian and a porch sitter could converse without shouting. In the days before television and air conditioning, the front porch was a crucial room and entertainment source (Jackson, 1985).

The streetcar neighborhood was never strictly residential. The main street, down which the trolleys ran, was usually lined with shops. This was the zone with the maximum of pedestrians, getting on and off the trolleys. These shopping streets ran like a spine through miles of residential neighborhoods, blocks of houses which began immediately behind the stores. By design, the homes were clustered within half a mile of the main boulevard. Thus, every home had stores in walking distance. This was valuable to the adults for their normal shopping needs – the equivalent, on a smaller scale, of the shopping centers and malls of the car suburbs. For children and youth, though, the streetcar suburbs were far superior to the car suburbs because there were a variety of interesting public places that kids could go on their own, without needing to be driven.

The greater physical density fosters greater social density. Dog walkers are a great source of community connection, and the porch sitters and dog walkers have regular conversations. Boburb residents have many reasons to walk – to the bus, the store, the coffeehouse, the park, or at least to their car parked on the street – giving them many opportunities to run into one another.

The trolley suburbs extended the city grid, connecting suburb and city in an easy and transparent way. At the time, this was not so much a design decision as simply the way things were done. Moreover, the straighter the trolley track, the easier and safer it was to manage, so the main boulevard, and all those streets parallel and cross-connected to it, tended to follow the grid, as well.

Boburbs draw parents who want to raise semi-urban kids – not in the "wild" bohemia, but not in the "sterile" suburbs, either. The ideal type of these families is educated parents who enjoy the cultural production of the city, the types of people who gather in its bars and coffeehouses. They want their children to experience more diversity of everything – people who look differently, act differently, and think differently from each other and from the norm. Such parents will know the city has risks, but also know that every place is safe if you know the rules.

The streetcar suburbs were originally designed for residents with no cars. They rode the streetcar to work, and walked to their other needs. As cars become more affordable for the middle class in the early 20th century, the design of new trolley suburbs came to incorporate small, detached garages, usually down a narrow driveway to the rear of the lot. Today, the streetcars are no more, and less frequent buses run up the old trolley boulevards. Many more cars squeeze into the streetcar suburbs than they were ever designed for, and parking is a constant issue.

Even today, though, the streetcar neighborhoods are about the most manageable of any kind of residences in America for those without a car or with fewer cars per household. Shops still line the main boulevard. Social gathering places – bars, coffeehouses, ice cream vendors, and every kind of restaurant – abound. The trolley suburbs were made in a great era of the religious Establishment, so churches are often incorporated into the residential streets. The trolley era was also the building era of public libraries, which constructed branches of the main downtown library amidst the new suburbs. Some public transportation still goes downtown. The core city is in bicycle distance, and traffic speed in this dense part of town is safer for cyclists. Moreover, a century or more of maturity has made an enviable tree cover, which improves the living – and walking – for residents and visitors alike.

The boburbs are usually the most diverse residential neighborhoods in the whole metropolitan area. The tolerance of diversity that is essential to bohemia carries over into the boburbs. Non-white middle-class people, moving out of their class and ethnic enclave neighborhoods, are more likely to choose a boburb than the whiter suburbs farther out. The boburbs are the great culture-making parts of a city region – the place of art galleries, music venues, art-house theaters, bookstores, non-degree educational offerings of all kinds, and the places that culture producers gather to talk. The coffeehouse is an appropriate symbol of the boburb.

The boburb is also the most politically liberal part of the metropolitan area, and, indeed, of the whole state. Liberal causes of all kinds are organized there. Liberal politicians can get elected there. The bulletin boards and telephone poles are covered with announcements of meetings, protests, vigils, rallies, and pure expressions of political sentiment – overwhelmingly left-of-center. The boburb is the "blue spot" in a red state, and the bluest spot in a blue state. Moreover, boburb residents are more likely to express their views publically than people in the car suburbs are. The boburb is the place to find cars and vans covered with stickers expressing ideological positions.

At the national level, the ideological preferences are clear: conservatives like the suburbs, liberals like the boburbs. The same boburb/suburb survey noted above found that the political spectrum neatly flows in the expected direction. On a scale from Very Conservative to Very Liberal, the suburb preference goes down, while the boburb preference goes up (Table 2.1).

Who do the boburbs exclude? Who do the boburbs displace?

The boburbs are not for everyone. Though they foster an ideal of inclusion and prize diversity, each boburb has displaced some kinds of people and excluded other kinds of people as it has come into its own. The core culture of the boburb is driven by educated, engaged, liberal, cultural omnivores (Peterson and Kern, 1996).

If the boburb grew out of a bohemia that gentrified, then the new class of residents displaced the poor artists who made the neighborhood interesting in the first place. In practice, artists who succeed economically, or who have middle-class day jobs or spouses, stay on as the neighborhood gets richer and gentrify themselves in place. If the preceding bohemia had itself grown up in a working-class or poor neighborhood, some of those earlier residents may live to see themselves priced out by bobos. The modal boburb, though, if built in an old streetcar suburb, has middle-class or even upper-middle bones to begin with. The boburb is, in a sense, a reclamation of a former class status.

Boburbs still tend to be very white, even though most residents I talked to wanted their neighborhood to be more racially diverse. The present racial mix reflects past "redlining" and racial exclusions, racial steering by real estate agents, and long memories of past exclusions by black and Hispanic families from other parts of the city. Today, white and non-white families are welcomed equally into boburbs, especially if they show the other signs of sharing in boburb culture.

Boburbs are notably liberal politically. Consequently, conservatives know they are at odds with the neighborhood norm, and sometimes feel positively

Table 2.1 Political ideology spectrum of suburbanites and boburbanites

Preference	Very conservative	Conservative	Moderate	Liberal	Very liberal
Suburb	61	58	46	36	22
Boburb	36	40	51	63	76
Missing/no answer	3	2	3	1	2
Total	100%	100%	100%	100%	100%

oppressed. In the boburbs, expressing your political views is also normal – in contrast to the suburban subdivisions. The conservatives I interviewed in the boburbs were mostly young people who planned to end up in the suburbs. The rare conservatives who stayed in the boburb to raise their families tended to be pro-life Catholics who relished their parish community, who walked to church and to the parish school, and were active in the life of both. They were bobos in all other respects besides their politics, enjoying the broad cultural options available where they lived.

The boburbs are uncomfortable for people who value privacy and keeping to oneself. In a boburb, the dog walkers, porch sitters, and people watchers keep an eye on, and a listening ear to, the neighbors as well as the visitors. By contrast, in the car suburbs, few people could even name all of their neighbors, much less tell us about them. The norm, rather, was that good suburbanites keep to themselves.

There are many fewer reasons to walk in the suburbs. The porches, if they exist, are too far from the sidewalks to speak to, and in newer suburbs there usually are no sidewalks at all. Sidewalks afford public (as opposed to intimate) relations with other people. In the sidewalkless suburbs, one can only have public relations with other people's things. Suburban dogs may not be walked regularly, but are more likely to be confined to the fenced backyard. One friendly couple in a car suburb sat on lawn chairs in their driveway, hoping to attract conversation from the passersby. Many lamented that even when they wanted to run into them, their neighbors drove into their automatic-opening garages and never came out. Social life in the suburbs is much more back-deck than front-porch – and by invitation only.

Boburbs exclude people who resent strangers coming through. The stores and public facilities create many reasons for visitors to come to the neighborhood. The road grid, through boulevards, and even the alleyways create dozens of paths to drive or walk. When the car suburbs were built, they made the opposite choice – winding roads, cul-de-sacs, bottleneck entrances – to make the car suburbs safer and greener (especially for children). This also makes the car suburbs much harder for strangers to navigate *by design*, and greatly reduces the diversity of kinds of people and functions one might meet there.

The core exclusion of the boburb, though, has been *class exclusion*. The main mechanism for this class sorting has been the sheer cost of buying a house in a neighborhood that consists mostly of single-family houses of medium to large size. However, the streetcar suburbs were built with a wide variety of housing types – much wider than the car suburbs have – so there have always been a few places in the boburbs for the poor and the rich to live.

A more subtle kind of exclusion – or, at least, differentiation – has been by "class fraction" (Bourdieu, 1984). College graduates who have middle-class or upper-middle-class occupations are part of the same social class, considered on a vertical ladder from poor to rich and from lower status to higher status. However, *within* the same class, there is often a difference of lifestyle, values, and even worldview between the fraction with more money, and the fraction with more culture. The "corporate class" fraction of business managers lives differently than the "knowledge class" fraction of culture producers, teachers, analysts, and interpreters. The corporate class favors the suburbs. Boburbia is the preferred home of the knowledge class (Weston, 2011).

Boburbia and its alternatives in Louisville

I asked the Centre College alumni the survey question we opened with:

> Imagine for a moment that you are moving to another community. Would you prefer to live in …
> 1. A community where the houses are larger and farther apart, but schools, stores, and restaurants are several miles away [OR]
> 2. A community where the houses are smaller and closer to each other, but schools, stores, and restaurants are within walking distance?

I then asked them which option best described the kind of neighborhood they actually live in.

The respondents in this sample, graduates of a selective liberal arts college, skew more toward the boburb than the US population as a whole does. Whereas the American population splits in equal halves given this choice, the Centre graduates favored the boburb 63% to 37%. When asked where they actually lived, though, the split was closer to even, with 56% in the boburb, 44% in the suburb. Comparing the actual suburban group with the actual boburban group is especially useful in seeing the true difference it makes to live in each kind of neighborhood.

First, consider how people in the two kinds of neighborhoods behave (Table 2.2).

One of the defining, appealing elements of a boburb is walkability. Boburbs are much more likely to have sidewalks in the first place, and boburbanites are more likely to walk regularly in any case. In private life, boburbs appeal to people who want more social interaction with their neighbors, which leads to having more friends in the neighborhood. In public life, boburbanites are more likely to be informed about, and interact with, their local government. The counterpart of the greater walkability of the boburb is the greater need to commute in the suburbs. It is almost a proverb that everything is "about a twenty-minute drive" in Louisville. When we look at their actual daily commutes, though, suburbanites are twice as likely to have a commute longer than that.

We can also compare how boburbanites and suburbanites feel about where they live (Table 2.3).

Boburbanites tend to be progressives; not surprisingly, they think the place they can affect is making progress. Suburbanites tend to favor control; they want their place to stay reliably the same.

The crime question turned out interestingly. Fear of crime is usually thought to be one of the big reasons people move to the suburbs. Indeed, we

Table 2.2 What Centre College alumni do in their neighborhoods

% of each kind of neighborhood*	Live suburb	Live boburb
My neighborhood has sidewalks	52	82
I walk regularly in my neighborhood	75	90
Some of my five best friends live in my neighborhood	24	40
I know the name of my Metro Council representative	40	55
I have met my Metro Council representative	20	31
My commute is greater than 20 minutes daily	38	19

Note: *All comparisons significant at the p<0.05 level.

Table 2.3 How Centre College alumni feel about their neighborhoods

% of each kind of neighborhood	Live suburb	Live boburb
My neighborhood is getting better	26	48
My neighborhood is staying the same	72	48
I worry about being a crime victim in my neighborhood*	10	20

Note: * Combines "Strongly Agree," "Agree," and "Somewhat Agree."

see that people in the boburb – who, you recall, very much like where they live – nonetheless worry more about being a victim of crime in their neighborhood. However, when I asked if they were worried about being a victim of crime in Louisville in general, the people who wanted to live in the suburbs were significantly more worried than the people who preferred a boburb. The suburban fear of urban crime is more a reflection of their worldview about "the city" than about their specific experiences of urban places.

As we see above, there are quite a few people living in the suburbs who would prefer the boburbs. When we compare the two groups based on their preferred kind of neighborhood, rather than their actual residence, some of the differences are revealing (Table 2.4). Those who prefer the suburbs are more conservative than the boburb-favorers, as we would expect.

Part of this political difference is related to stage in the life cycle – boburbanites, as a group, tend to be young, single, and childless, whereas the suburbanites include more older, married parents. Still, as we can see from the preferences, there is an ideological difference between the two groups beyond the effect of their life stage. Liberals prefer social density. Conservatives prefer social control. This very much affects where they want to live.

Here we are heading into the deep waters of why people have different views of the world to begin with. One of the standard social science tools to study this issue is with the trust question, which has been asked in surveys for decades (Table 2.5). The usual wording is this: "In general, do you think that

Table 2.4 Centre College alumni political ideology, by neighborhood type

% of each kind of neighborhood	Like suburb	Like boburb
Very conservative	10	2
Somewhat conservative	30	8
Moderate	27	19
Somewhat liberal	20	31
Very liberal	13	40

Table 2.5 Centre College alumni trust level, by neighborhood type

% of each kind of neighborhood	Like suburb	Like boburb
Can be trusted	64	81
Can't be too careful	37	19

people can be trusted, or alternatively, that you can't be too careful when dealing with people?" (Uslaner, 2012, 72ff).

Note that in this privileged group – educated, higher-class, white people living in a US city – most people are trusting. Still, those who want to live close to others are overwhelmingly a trusting group. And this despite the fact that boburbanites have a higher (and realistic) fear of crime in their neighborhood. Trust and social density go together.

A central question of this study is what kind of *families* favor the boburb. Parents are more involved in their communities than non-parents across the board. But boburban parents are just a bit more involved in their communities than suburban parents are. Comparing just boburban parents with suburban parents lets us make an apples-to-apples analysis.

Two-thirds of parents who actually live in the suburbs prefer the suburbs – but almost 90% of boburban parents prefer the boburbs. Raising children in the suburbs is the default path for Americans; raising kids in the boburb is a more deliberate *choice*. Most suburban parents like their specific neighborhood a great deal (76%), but boburban parents like their neighborhood just a bit more (81%).

In the suburbs, having kids changes people into neighborhood walkers (rising from 58% to 82%). Boburbanites were walkers already (89%) – but parenthood pushes them out into the community even more (91%). Compared to suburban parents, boburban parents are a little more likely to have best friends in the neighborhood (31% to 26%) and know the names of all their neighbors (87% to 79%). Boburban parents are notably more likely to be involved in local government, both in knowing who their Metro Council representative is (69% to 45%) and in having met that person (47% to 22%). Parenthood tends to make people more religiously active, but has a slightly bigger effect on boburban parents. In both places, only half of non-parents are involved in a religious organization. When they become parents, this proportion rises to 65% in the suburbs, and to 71% in the boburbs. Boburban parents are also a little more likely to be involved in other face-to-face organizations than their suburban counterparts (62% to 57%).

Parenthood seems to improve how boburbanites experience their neighborhood. While only 11% of non-parents in the boburb strongly believe that "People around here are willing to help their neighbors," 31% of boburban parents think so. The great project of the suburbs is raising children, so we might expect parents there to feel real solidarity with one another. A bare majority of suburban parents do, indeed, think their neighborhood is "close-knit" (57%) – but significantly more boburban parents feel that way about where they live (73%).

We can get a first taste of boburbia and its most potent alternative, suburbia, from a few examples. All of the Highlands parents said they valued the "diversity" of the neighborhood. They were quick to acknowledge that it is not very racially diverse, but they relished the range of eccentric people – the tattoos, piercings, "green-haired kids," as well as the range of ages and sexual orientations. They wanted their kids to have some independence, to navigate public transportation on their own, hang out with other kids without constant supervision, and learn some street smarts. As one Highlands parent put it, "I trust my kids – and I want them to have some freedom now, so they don't go buck wild in college." To "go Bardstown Roading" was a constant pastime of Highlands youth – and of those suburban kids fortunate enough to get their parents to bring them into town for a time.

The experiences of two families are instructive. Both were white, married, multi-child, upper-middle-class business families. They became friends when living in similar upper-middle-class suburban subdivisions. When their children were high schoolers, both families decided to change neighborhoods. One built a large house (over 6,000 square feet) on a multi-acre lot in a gated community in the far reaches of the county. They sent their three children to a suburban religious school. They were very active in their megachurch, and often hosted their children's youth groups and sports teams. They are conservative, religiously and politically. They had a nodding acquaintance with several of their neighbors, but did not socialize with them.

The other couple moved to the Highlands, to a 2,400 square-foot house in walking distance of Bardstown Road. Their two children went to different public schools, drawn by distinctive magnet programs. Of the Highlands, they enthused that they "loved it from day one." They liked the big trees, and the smaller yard. They were "big foodies" and enjoyed walking to non-chain, locally owned restaurants. They like the diversity of age, lifestyle, and ethnicity, and want their children to see it – that is also why they sent their kids to public schools. They met other dog walkers, joined a local book club, had regular happy hours with the other residents, and joined a local, walking-distance (liberal) church. "We knew our neighbors better within a year," they said, "than we did in six years" in their old neighborhood.

References

Pierre Bourdieu, 1984. *Distinction: The Social Critique of the Judgement of Taste.* Cambridge: Harvard University Press. Translated by Richard Nice.

David Brooks, 2000. *Bobos in Paradise: The New Upper Class and How They Got There.* New York: Simon and Schuster.

Emile Durkheim, 1893. *The Division of Labor in Society.* Translated by W.D. Halls. New York: Free Press, 2014.

Kenneth Jackson, 1985. *Crabgrass Frontier: The Suburbanization of the United States.* New York: Oxford University Press.

Richard Peterson and Roger Kern, 1996. "Changing Highbrow Taste: From Snob to Omnivore." *American Sociological Review,* 61, 5: 900–907.

Pew Political Polarization Survey. 2014. www.people-press.org/2014/06/12/political-polarization-detailed-tables/.

Eric Uslaner, 2012. "Measuring Generalized Trust: In Defense of the 'Standard' Question." In Fergus Lyon, Guido Möllering, and Mark N.K. Saunders, eds., *Handbook on Research Methods on Trust,* second edition. Northampton, MA: Edward Elgar, 72ff.

William Weston, 2011. "The College Class at Work and at Home." *Society,* 48, 3: 236–241.

3 How did the Highlands happen?

From suburbia to bohemia to boburbia

The Highlands were originally just hillsides on the edge of the city. As a neighborhood, they were born as an outer suburb of Louisville, were transformed into an inner suburb, did a short turn as a bohemian quarter, and have now become a boburb. This is, no doubt, not the end of the Highlands Story.

The first fifty years: founding the Falls City, 1778–1830

The Ohio River was the highway into the heart of the continent in colonial America. King George III tried to keep the Americans from following that highway. He declared a Proclamation Line of 1763 along the continental divide in the Appalachian mountain range as the western boundary of colonial expansion. When the Americans began a revolution against the king, that line became dead letter. General George Washington sent Colonel George Rogers Clark to establish a base on the Ohio for military operations against the British and their Native American allies north of the river (Yater, 1979).

In the thousand-mile length of the Ohio, the river descends gently from its origin in Pittsburgh to its merger with the Mississippi River at Cairo, Illinois. In all that length, there is only one serious obstacle to boat traffic, the Falls of the Ohio. Six hundred miles down from Pittsburgh, a series of rocky ridges rake the river at low water. Clark saw the strategic importance of this pinch-point on the river. Civilians, eager to develop the interior, saw the commercial possibilities of an essential portage point for goods traveling on the river.

In 1778 Clark arrived with 150 militiamen and 80 settlers. They first set up on an island in the falls, then established "Fort on Shore." From this point they slowly built two settlements. To the east, along the river, was the port of Louisville, to meet boats coming down from Pittsburgh. To the west, on the other side of the falls, was Portland, to send boats down the Ohio toward New Orleans. The economic life of the two cities would, for their first fifty years, focus on connecting those two waterfronts.

During the American Revolution, Clark successfully raided British and Native American forts to the north. The British fought back, and often retook what they had lost, but they could not keep defending these far outposts of their empire. The Indians mostly stayed out of the war with the

Americans, seeing that the British were not reliably powerful allies. In 1780, the revolutionary government of Virginia, which claimed the Kentucky territory, helped court French help in the war by naming this new settlement Louisville, in honor of King Louis XVI. At the same time, they named the county for Virginia's governor, Thomas Jefferson.

The Americans won the war, and were quick to expand beyond the mountains. In 1792, Kentucky became the first western state.

Kentucky happened when and where it did partly because of who it did *not* displace. Surprisingly, there were no Native American villages in Kentucky when the American settlers came (Henderson and Pollack, 2012, 408). The Shawnee to the north, and the Cherokee to the south, preserved Kentucky as a hunting ground and neutral zone. The Shawnee were being pressed by Iroquois expansion from New York. The Cherokee were beginning their long, painful encounter with the Americans in the South. The Chickasaw lived on the Mississippi River far to the west. And all the tribes had been devastated by first contact with European diseases, especially smallpox. Each of these tribes would be commemorated in a set of significant parks in Louisville a century later. But none were displaced by the settlement at the Falls of the Ohio. The Americans, coming down the river from Pennsylvania or, like Daniel Boone, up through the Cumberland Gap in the mountains from Virginia, filled in the unsettled (though not unused) land between the native nations.

Kentucky was born with a serious exclusion, though: the exclusion of African Americans from full citizenship due to slavery. Virginia was a slave state, so its laws came to Kentucky with the settlers. African Americans built Louisville from the beginning. There were even free blacks there from an early day. But for the first hundred years, black Louisvillians were brutally excluded, and, for a hundred years after that, were severely constrained. The city's third century looks to be much better in that regard, but the struggle, a generation into this new era, is far from over.

Louisville grew slowly, creating a few streets parallel to the Ohio River. Commercial building filled in the muddy space between the river and Main Street. The next long east–west street was Market, and the next was Jefferson. The fully built-up city was only twenty blocks along these three principal streets, starting from 12th Street on the west end, where Fort on Shore had been built. South of Jefferson the lots got bigger, with houses and grounds that still had significant gardens and even livestock. Walnut and Chestnut Streets had half the density of the first three. The southern boundary of the city, Prather Street, opened on to farms and open country. This road was wider than the other city streets, a pattern that would continue with the outer perimeter roads several more times in the expansion of Louisville. The perimeter road would also be a boon to the suburbs of the city, connecting them with one another without requiring a trip in to the center of town.

Louisville, like all cities from the ancient *polis* to the early 1800s, was a walking city. While the rich came to have horses and carriages, for the mass of people, a walking radius was the limit of urban development. Most walking cities concentrated everything – making, buying, selling, governing, playing, and living – within a two mile radius of the "city hall" or whatever was the town center (Warner, 1962; Jackson, 1985). Prather Street, the boundary of early Louisville, was about a mile and half from the river's edge.

The economic leaders of Louisville could see that the city would reap a new fortune if it could solve two problems: getting boats around the falls, and getting boats to go upstream economically. In the 1830s they took a risk to solve the first problem: they built a canal around the falls. Immediately, the economic niche changed for Portland, and the adjacent little village whose name tells its function – Shippingport. The second problem was solved at the same time when economical steamboats were perfected that could come up the Mississippi and the Ohio. Louisville boomed with flowing two-way river traffic. The city would grow west along the river, absorbing Shippingport and Portland in the 1850s (Rogers, 1955, 4).

The second fifty years: inventing the Highlands, 1830s to 1880s

Louisville burst its village boundary. Prather Street, the first perimeter parkway, became the descriptively named Broadway. Bigger houses were built on Broadway, as new streets were laid out to the south (Rogers, 1955, 36). Mulecars, and then a steam train, connected Louisville and Portland. The city of Louisville was incorporated and started making public schools. In the 1840s the train line connected Louisville with Frankfort, the state capital. In the 1850s the Louisville and Nashville Railroad, the dominant corporation of the city for decades, completed its central line.

The political life of the city was shaped by White Anglo-Saxon Protestant (WASP) slaveholders. They controlled the various political parties to maintain their hold on the city, which was challenged as the ethnic composition of the city changed. In the 1840s, significant numbers of Germans and Irish immigrated to Louisville. The Germans were generally Protestant, but of denominations not previously found in the city, as well as some Catholics, and the city's first Jewish congregation. The Irish were overwhelmingly Catholic. Some of the Germans were socialist "48ers" who fled the failed revolutions in their homeland. By the 1850s, Jefferson County was 17% German and 7% Irish immigrants, concentrated in the working-class neighborhoods surrounding downtown. These urban immigrants were not big supporters of slavery, for reasons of both moral objection and economic competition. African Americans constituted another 16% of the city, three-quarters of whom were enslaved (Hudson, 2011).

The conflict between the pro-slavery, anti-immigrant mobs and the German and Irish minorities led to the bloodiest riot in Louisville's history, the 1855 Know-Nothing Riots. The Know-Nothings were a short-lived political party, explicitly anti-immigrant and anti-Catholic, and implicitly pro-slavery. On election day, August 6, Know-Nothing patrols "protecting" the polls from Democrats turned into a mob. They attacked immigrants, especially in

German Butchertown. Though 22 people were killed, many injured, and much Irish and German property was destroyed, no rioters were convicted and the victims were never compensated (Ullrich and Ullrich, 2015).

In the Civil War, Louisville remained in Union hands, and was a vital center of war industries and military supplies. Louisville, and Kentucky as a whole, were pro-Union, though not necessarily anti-slavery. The Emancipation Proclamation, which freed slaves in the Confederacy, did not apply to Kentucky. Slavery was not abolished in Kentucky until the passage of the 13th Amendment to the Constitution. The power structure of the city had always been dominated by slaveholders and businessmen who benefited from slave production, even if they did want the union to stay together. After the war, therefore, Kentucky became the only place to "join the losing side after a civil war," as the old political quip held. Former Confederates quickly came to dominate government, in Louisville as well as in the state as a whole (Marshall, 2010).

After the war, Louisville was one of the few southern cities that had not been ravaged by the fighting. It became the crucial Gateway to the South, especially by way of its railroad network, and boomed as an industrial city. In 1870 the first bridge over the Ohio River was built, for the railroad (Yater, 1979, 141). Two signature businesses developed in this period: in 1875 the first Kentucky Derby was run, and in 1883 the original Louisville Slugger baseball bat was produced. Louisville took a significant step into cultural leadership, as well, when in 1877 it lured the flagship Southern Baptist Theological Seminary away from Greenville, South Carolina. 1880 was a turning point for Louisville migration. Before that, it was just another stop for a transient population. After that it was an industrial and commercial city of some consequence, the natural magnet for talent from its hinterland (Bower, 2016, 80).

With that boom came the first suburbs.

The third fifty years: completing the Highlands, 1880s–1930s

The first planned "suburb" to grow directly contiguous to downtown Louisville came from the Southern Exposition. This was Louisville's big public show of its industrial strength. Held each year from 1883 to 1887, the exposition grounds extended south from the city, into then undeveloped countryside. After the exposition, the promenade was turned into Central Park. At the south end of the park a remarkable set of New York-style brownstones was developed on Belgravia Court and St. James Court, now home to a large annual art festival. The numbered streets were extended southward, with a grid of connecting streets in between. This region drew the mansions of the rich away from Main Street, which was increasingly commercial, into this new fully residential suburb now known as Old Louisville.

The working-class neighborhoods were crowded around the central town, in walking distance of river-based work. African Americans, enslaved and free, were joined in the 1840s by significant numbers of Germans and Irish. Each lived in separate enclaves, with ethnically identified names like

Smoketown, Limerick, Irish Hill, and Germantown, along the southern edge of the city. These neighborhoods were not so much planned suburbs as self-created housing for groups not welcome elsewhere in town, as close to work as they could get.

With the coming of the horse-cars and mule-cars on fixed tracks, however, it became economically viable to build planned suburbs for the middle classes and at least the higher end of the working class. These animal-powered streetcars extended the reach of a viable worker's commute to a radius of six miles from the city hall. Tracks were laid for mule-drawn cars west to the bend in the river, east along the road to Frankfort, southeast on the road to Bardstown, and south to the farmlands open to development. By 1890 there were 125 miles of streetcar tracks in Louisville (English, 1972, 40). These streetcar suburbs greatly expanded the living range of middle-class and secure working-class families. When the mule carts were replaced by electric streetcar lines in the 1890s, there was a boom in suburban house construction for the middle class and most secure working class (Jackson, 1985).

The Frankfort Avenue and Bardstown Road lines would become particularly important. They were laid out on either side of a vital amenity of a Victorian city – the cemetery. Cave Hill Cemetery (established 1846) was, in effect, the public park on the outskirts of Louisville. It was common in the 19th century to make family outings to cemeteries, to tend the graves of family, and to enjoy a bucolic picnic and ramble.

The extension of the Broadway streetcar line to the gates of Cave Hill Cemetery is the true beginning of the Highlands neighborhood.

To see why, it helps to know why the height of the Highlands matters. Louisville's physical development was shaped by the microgeologies of drainage. Built on a limestone plateau, the whole territory is riddle with karsts – the hollowed-out places where soft rock dissolves around harder stone. In addition, like all river cities, Louisville was subject to occasional floods. Thus, as the city spread away from the riverfront, the most desirable territory for the richer classes to develop was in the higher ground located nearest the city center. One line of early development grew along the ridge heading southeast from downtown that would give the Highlands their name. Another line of development headed east along the river, skipping the low, malarial marsh of Beargrass Creek, and taking up again on the higher ground that leads to Crescent Hill. This high ground was sought after by richer people for their homes, which in turn drew businesses catering to the middle classes and above (Conkin, 2003).

The Highlands as a neighborhood began as a fancy suburb on the edge of town. The big houses were advertised as offering the benefits of a cleaner, quieter, more natural life for the family, while being close to business and the amenities of the city. When that section started to develop a real identity, the Highlands name became fashionable. This, in turn, led adjacent neighborhoods to be developed, expanding the Highlands umbrella. The point at the edge of town from which all these developments sprang, where Broadway reached the cemetery, was retroactively christened the Original Highlands. This is a pattern we see often in the development of suburbs – they are first sold to the upper-middles as the best of country and city living, then draw the larger middle class, the businesses that serve them, and the schools, religious institutions, and other socializing institutions which those classes animate.

The first great residential development of the Highlands was in what became known much later as the Cherokee Triangle. James Henning and Joshua Speed were already successful businessmen and active citizens when they planned the first part of the "Henning and Speed Addition." This spot was appealing in several ways. It was on higher ground, it abutted the city's park-like cemetery, was within the city services area, and had just been reached by the streetcar line running down Broadway, the former perimeter road. This first section of the Highlands was so appealing that Henning and Speed planned to build their own houses there.

The right-hand (eastern) boundary of the triangle was Cave Hill Cemetery. The left-hand boundary was Bardstown Pike, a dusty, unpaved track down which farmers brought cattle from rural Kentucky. Down the middle of the triangle they proposed a new, wide street, beginning at the eastern end of Broadway, which they hopefully named East Broadway. This major road was crossed by a connecting road, aptly called Transit Street, which took travellers from Bardstown Pike and the growing southern reaches of the city, over east and north around the back of the cemetery, to the Lexington Road. At the corner of East Broadway and Transit Street, Henning and Speed built their first mansion, a wedding present to Henning's daughter.

The appeal of the cemetery led to a movement to create true public parks. In the 1890s, Louisville commissioned the greatest landscape architect of the day, Frederick Law Olmsted, to design a series of parks around the edge of the city, named for Native American tribes. The 19th-century streetcar lines extended out in spokes from downtown to these parks. The western line went to Shawnee Park, the southern to Iroquois Park, and the southeast and eastern lines ran on either side of a string of green spaces, Cave Hill Cemetery, Cherokee Park, and, later on, Seneca Park. The southeastern line, down Bardstown Road, became the core of the Highlands, and the eastern line, down Frankfort Avenue, became the core of the Clifton and Crescent Hill neighborhoods.

The opening of Cherokee Park led the Longest brothers to build the next subdivision south of the Henning and Speed Addition. To capitalize on the park, the Longests built a triumphal boulevard, Cherokee Parkway, leading into the park. They then lobbied to have a streetcar line built down this road, enhancing both the park, and their development (English, 1972, 103).

The Olmsted parks were joined together by the new version of perimeter roads around the city. A system of parkways connected these parks, about four miles out from the center of town. They also extended the city outward, and contributed to the development of what today have become the inner suburbs.

Figure 3.1 The incomplete system of Olmsted Parkways, shown with the interstate highways (used with permission of Olmsted Parks Conservancy)

The Cherokee Triangle was built by developers and land speculators. They planned for big houses of durable material on the long streets, and sold lots for owner-built frame houses on the crossing streets. They encouraged the leading Protestant denominations to build on East Broadway – now Cherokee Road – in the midst of the residential neighborhood. They lobbied to get the trolley lines extended to and through their developments. They donated broad streets to carry the trolley lines to the gates of the parks. They encouraged stores along the adjacent Bardstown Pike, and packed dense blocks of potential shoppers and commuters into the new neighborhoods behind.

The most careful study of the Cherokee Triangle found that most of the people who moved there were not city people moving out, but country people moving in. In 1910 the few Germans and Irish who moved up, socially, to the Cherokee Triangle grew up in the city. However, nearly all the rest of this tony suburb's original residents grew up in the county-seat towns or city-oriented farms. They had enough connection to the urban economy to come to the city with white-collar skills and aspirations. The Cherokee Triangle was about as green as a dense urban settlement could be. It had been carved from farms, and was still connected to the country at the southeast end. The whole north and east side would be lined with the parks of Cave Hill Cemetery, Cherokee

Park, and Seneca Park. The Queen Anne style that most chose to build their houses in made the houses irregular, like a farmhouse. Highland Presbyterian's gothic style looked like a country church, as did the Episcopalian Church of the Advent (Bower, 2016, 159).

This outer suburb of its day served a function we see again and again with the outer suburbs: they house people who want city jobs and city incomes, but not the dense, dirty, noisy, scary parts of urban life. The outer suburbs are a compromise social destination for people moving toward the city. This is just what the non-city people want in a country club suburb today.

By the turn of the century, the broad avenue running down the center of the Cherokee Triangle, originally East Broadway and now Cherokee Road, was a paved street with a streetcar line. It had four active churches on it, the Episcopalian Church of the Advent, and three bearing the Highlands name, Highland Presbyterian, Highland Baptist, and Highland Methodist. Dusty Bardstown Pike had turned into paved Bardstown Road, with its own streetcar lines. Bardstown Road was lined with small stores, the commercial strip that served the expanding residential neighborhoods of Cherokee Triangle to the east, and the new neighborhoods being built to the west.

The successful development of Cherokee Park and the Cherokee Triangle on the east side of Bardstown Road naturally led to interest in developing the west side. Tyler Park grew westward beginning in the 1880s, with the subdivisions getting fancier and more expensive up to 1910. The most "suburban" part of the Highlands, the Castlewood section, was designed by the Olmsted firm with winding roads and cul-de-sacs, unlike the rest of the city grid (Sachs, 1997). The surrounding area was then so much on the outskirts of town that four cemeteries were established there – St. Michael's for German Catholics (1851), St. Louis, in its third location, for Irish Catholics (1867), and Louisville (1886) for African Americans. In 1921 the Roman Catholic Archdiocese would build its largest cemetery, Calvary, just outside of Tyler Park. The neighborhood takes its name from a city park, named for a recently deceased mayor, that opened in 1910. The neighborhood is bounded on the south by Eastern Parkway.

The outer edge of Louisville had passed on beyond the Cherokee Triangle. Eastern Parkway, part of the new perimeter road which connected the "string of emeralds" that were the new Olmsted Parks, formed the southern edge. As always, the perimeter road just spurred new development on it and beyond it.

On the east side of Bardstown Road, and continuing south of the Cherokee Triangle, is Bonnycastle. The Bonnycastle addition was developed from the 1890s to World War I. This was the new upper-middle-class outer suburb, framed by Eastern Parkway on the north, Bardstown Road, and Cherokee Park. In this neighborhood of single-family homes, apartment buildings were rare. However, in 1928, one notable exception was made: the Commodore Apartments. This 11-story tower, overlooking the park, was designed, built, and owned by leading Jewish families, and became a choice residence of Jewish Louisville in the Highlands. Two synagogues – Adath Jeshurun and Keneseth Israel – are in walking distance (Ely, 2003, 162).

The working-class and lower-middle-class counterpart of Bonnycastle, west of Bardstown Road and south of Eastern Parkway, was Deer Park. In the 1890s, rows of long, narrow shotgun houses were common; in the early 20th century construction fashions shifted to bungalows. This neighborhood was especially defined by the Bardstown Road streetcar line, which was extended in 1912 to the turn-around at Douglass Loop. The loop marks the southern boundary of the neighborhood.

By 1900, Louisville had added more that 80,000 people in just 20 years. Much of this growth was heading out Bardstown Road. The new suburbs were built on each other continuously. By World War I the last extension of the streetcar had been made, and the last Highlands neighborhoods in walking distance of Douglass Loop were being filled in.

On the east side of Bardstown Road, extending over to Seneca Park, is the Highlands Douglass neighborhood. Most houses were built after World War I. The fancier ones, closer to Bonnycastle, were built before the Depression, while the outer portions of the neighborhood were not finished until the 1950s. Seneca Park, the last of the Olmsted Parks, was built on land confiscated from a German aristocrat who fought for the Kaiser. The park was officially opened in 1928, and is the only one to be built with a course for the fashionable suburban sport of golf.

Highlands Douglass' counterpart on the west side of Bardstown Road is Belknap. After World War I there was a movement to consolidate the various pieces of the University of Louisville here, in a territory to be called University Park. Several street names were changed to Harvard, Yale, and Princeton in anticipation of the development. However, the bond issue to consolidate the university failed. In 1925 a second bond issue succeeded when the University and the city agreed to create Louisville Municipal College for African Americans in the west end. The University of Louisville then consolidated its facilities for white students in Old Louisville. A generation later, when the law preventing integrated education was overturned, Louisville Municipal College was absorbed by the University of Louisville (Yater, 1979, 184, 219).

Decades after University Park was first imagined, in the 1950s, the Roman Catholic Archdiocese would found Bellarmine University in the Belknap neighborhood. Another landmark of this neighborhood is the Lakeside Swim Club, begun in 1924 in a former quarry. Swim clubs have become among the most attractive amenities a suburb can have.

The Upper Highlands, the last neighborhoods to claim the Highlands brand, were largely created after World War I. They continued to grow until the Depression slowed all development to a crawl.

East of Bardstown Road is the neighborhood of Hawthorne, and four microscopic sixth-class "cities" – Strathmoor Village, Seneca Gardens, Kingsley, and Wellington. The Strathmoor neighborhood, the next area south of Highlands Douglass, was developed into several parts in the 1920s. In 1928 Strathmoor Village incorporated as its own small city of about 600 people, in part to resist annexation by Louisville. Seneca Gardens, also developed in the 1920s, was refused annexation by Louisville during the Depression, so incorporated as its own city, now numbering about 700, in 1939. Immediately to the east of Strathmoor Village, Kingsley, a subdivision of 400 people, incorporated as a city in 1939. Hawthorne, the neighborhood south of Strathmoor and bounded on the east by Louisville's original airport, Bowman Field, was also developed in the 1920s. In 1946, Wellington, another tiny city of 500 people, carved out a few blocks of Hawthorne for similar home-rule reasons.

West of Bardstown Road are two adjacent neighborhoods, Hayfield Dundee and Gardiner Lane, and the micro city of Strathmoor Manor. Strathmoor Manor, just across Bardstown Road from Strathmoor Village, is an entirely residential strip of about 300 people. In the southwest corner of the Upper Highlands is Hayfield Dundee. Its story as a Highlands neighborhood does not really begin until 1962, when Atherton High School (1924) moved there from the Original Highlands. Gardiner Lane, adjacent to Bardstown Road, began building subdivisions in 1913, though the process was not completed until the 1960s.

Between the turn of the century and 1930, another 100,000 people had been added to Louisville, bringing the city's total over 300,000. The Highlands had led the way in suburban development, carrying the city through the streetcar years and to the beginning of the automobile age. In 1920, 90% of the county population lived within city limits, and there were only 16,000 cars. By 1930, there were 50,000 cars, one for every seven residents. The new subdivisions were no longer built close to trolley lines, because the residents now owned their own automobiles (Yater, 1984, 17).

With the completion of the Highlands, Louisville's high-end suburban development shifted to the east end.

The fourth fifty years: the Highlands flourish and decline, 1930–1980

After the Depression and the war, the Highlands flourished in the 1950s as a mature inner suburb. After the race riots in the late 1960s, though, and the boom of the automobile suburbs going further and further out into the county, the Highlands declined. These conditions gave the Highlands a new life as Louisville's bohemia.

Almost no one was displaced as the Highlands were constructed, as each neighborhood was constructed on farmland or reclaimed marsh. However, the Highlands were built on exclusions. First and foremost, African Americans were excluded from nearly every part of the Highlands. Louisville was the site of a crucial Supreme Court decision, *Buchanan v. Warley* (1917), which outlawed the government engaging in direct racial exclusion in housing. This led to the indirect practice of redlining, whereby the government made racially tinged maps of neighborhoods assessing good and bad investment risks, but left it up to private mortgage bankers and real estate agents to apply them in a highly discriminatory way. The black west end was starved for investment, while the white Highlands was A-rated and green-lined. With the improvements in transportation, the African-American servants who used to live in the alleys near the Highlands houses they served could now be pushed out to live elsewhere and commute in.

There were exclusions among "whites," too – though not all would have been considered the same "race" at the time. The Cherokee Triangle was still heavily Protestant, though some Catholic families lived there. Still, no Catholic Church opened on the richer eastern side of Bardstown Road – St. Brigid (1873) and St. James (1901) are conspicuously on the western side of the divide. One measure of progress in overcoming these old divisions, though, is that when the first city planning commission was created in 1929, it was headed by an Irish-American named Murphy, and assisted by a German-American city engineer named Krieger (City Planning Commission, 1929, 10).

Jews were also explicitly excluded by deed covenants on some streets until after World War II. After the war the restrictions among white people would dissipate. Jews, in particular, would make the Highlands the center of Jewish life for a golden moment in the childhood of Baby Boomers. This fact would be important in the subsequent cultural flowering of the Highlands. By 1960, more than half of Louisville's Jews lived in the Highlands or a few other east end neighborhoods, and four of the city's five synagogues had moved to the Highlands. Jews contributed disproportionately to liberal movements in the city, especially to the white participants in the Civil Rights Movement (Ely, 2003, 153).

On the other hand, after the race riots of 1968, white people fled all parts of the city, not just the west end neighborhood where the conflict was worst. The school busing plan of the early 1970s greatly accelerated white departures from the city schools out to the Jefferson County school district (K'Meyer, 2009, 187ff). Black people did not move into the Highlands in any numbers, even when, after the Fair Housing Act in 1968, they could. The Highlands today remain more than 90% white.

By 1930 the Highlands developed a clear identity as a string of white, middle-class neighborhoods along Bardstown Road. Bardstown Road itself had developed as a vital commercial strip of small shops. While Louisville's leading department stores were still downtown on Fourth Street, the everyday needs of residents in the streetcar suburbs were met by the local stores along their commuting route, or in stroller distance of their homes. The older sections were maturing into vibrant residential neighborhoods of economically secure families. The newer parts of the outer Highlands were in the first decade of their full community life.

The Highlands just missed becoming the academic center of Louisville, losing out to rival Old Louisville, then a fading mansion district. As noted, the movement to pass a bond issue to create a campus for the University of Louisville in the outer Highlands failed in 1920. The city and the philanthropic Belknap family then stepped in to create the current main campus of the university at the southern end of Old Louisville.

The 1920s were the era in which one of the most distinctive features of native Louisville culture was instituted – "Where did you go to high school?" This question, ubiquitous among natives, puzzling to newcomers, tells a great deal about race, religion, and class, as well as neighborhood.

In the 19th century there were two parallel school systems, one public and dominated by Protestants, the other Catholic. While there was an attempt to begin the public system as early as the 1820s, the oldest high school in Louisville today is Presentation Academy, a Catholic girls schools founded in 1831. It began in what was at the time the Catholic quarter at the edge of the original city, on 4th Street near the present Cathedral of the Assumption. In the 1880s, Presentation would move to the new suburbs south of Broadway in the Irish Catholic neighborhood of Limerick, where it still exists.

The public high schools begin in earnest with the descriptively named Male and Female in 1856, followed by Central Colored in 1885, and completing the set with DuPont Manual in 1892. These schools leapfrog southward with the course of suburban development. Male and Female begin in the old city north of Broadway. By the turn of the century they were in or near Old Louisville, just south of Broadway. In the 1950s the public schools became coeducational, Male by adding girls (while keeping the name), Manual and Female (now Girls) by merging under the Manual name. The enlarged Manual moved further south, to the far end of Old Louisville, adjacent to the University of Louisville. Male remained at the north end of Old Louisville until the 1990s, when it took over a suburban high school building miles south of the Highlands, outside the Watterson.

Central (which eventually dropped Colored from the name, though not from its primary constituency) actually began south of Broadway in Limerick, but then moved back to the older part of town to 9th Street, the customary dividing line between the black west end and downtown. The other main high school to serve the west end, Shawnee, was built in 1929, as part of the maturing streetcar suburb on that side of town.

The Catholic schools are a stronger part of Louisville education than they are in most Southern cities, reflecting the city's location at the upper edge of the Protestant South. The early Catholic schools were made with less of a plan than the public schools, with different orders filling the need as they could. The Catholic high schools have remained single-sex. For girls, Presentation (1831) was created by the Sisters of Charity of Nazareth, Sacred Heart (1877) by the Ursuline Sisters, and Mercy (1901) by the Sisters of Mercy. Sacred Heart was established on the outskirts of town, in what would become Crescent Hill, where it remains. Mercy was created on the edge of the Original Highlands, and remained there until it moved to the far suburbs, outside the Watterson, in the 2000s. For boys, St. Xavier, by the Xavieran Brothers, began in the Catholic corner near the cathedral, but moved to the northern edge of Old Louisville, right on Broadway, in 1900. They remained there until the 1960s, when they leaped to a much larger campus south of the Eastern Parkway. In the 1950s the Archdiocese created a new school for boys,

Trinity, in the eastern suburb of St. Matthews, and asked the Sisters of Mercy to create a second girls school, Assumption, in the Tyler Park section of the Highlands.

Two high schools serve the Highlands, one public, and the other a secular, private preparatory school. The public school, Atherton, was built in 1924 in the Original Highlands, at first as a girls school. In 1915 the Speed family, who had made some of their money creating the Highlands, created Louisville Collegiate School for girls in an Old Louisville mansion. In 1927 they moved Collegiate to the Cherokee Triangle, where it remains. The school added boys in 1972. As the name suggests, it has been a college preparatory school from the beginning, and well reflects the aspirations of the upper-middle-class mature suburb that the Cherokee Triangle had reached at that moment. It had an informal counterpart in the Rugby University School for boys, which was founded in 1872 downtown, and would finish its days in 1949 in the Highlands Douglass neighborhood.

The enclosure of the Highlands was sealed by the next perimeter road, today's Watterson expressway. Indeed, this road began at Bardstown Road, with the first section connecting it to Breckenridge Lane two miles to the east in 1949. This two-lane "Inner-Belt" was overwhelmed by suburban growth in the 1950s. In 1958 the road was designated Interstate 264 to qualify for federal funds. The limited-access highway, now renamed for Henry Watterson, the 19th-century editor of the *Courier-Journal* newspaper, made a big half-circle south around the city. It connected to Interstate 64 just outside of St. Matthews on the east, to just outside Portland on the west. In the 1960s the Watterson was extended north to connect with Interstate 71, which runs from Louisville northeast to Cincinnati. The Watterson contained the whole city of Louisville (with a few exceptions on the west end). Eventually it would be thought of as the informal boundary between the city and the car-based suburbs. The Upper Highlands reach to the Watterson. The Frankfort Avenue boburb ends at the Watterson at Mall St. Matthews, the first enclosed shopping mall in Kentucky, opened in 1962.

From the end of World War II to the end of the 1960s was a golden age for the Highlands. The subdivisions filled in the hilly farmland east of Bardstown Road, and the swampy land to the west. The interurban trains ended in the 1930s, and the streetcars died out – the last run was in 1948, to be replaced by buses (Yater, 1979, 204). This was a sign of increasing wealth in the middle class, who could now afford their own automobiles. The 1937 flood showed the value of the Highlands as high land. 4th Street, downtown, was still the core shopping area for the city, and the mom-and-pop shops along Bardstown Road were healthy.

Then the suburbs started pulling. The car suburbs offered more house and a bit of ground for less money. The cul-de-sac subdivisions seemed safer for kids than the high-traffic city. The interstate routes in to the city – I64 from the east and I65 from the south, and the easy connection between them on I264, the Watterson – meant car access to most places in the city took "about

20 minutes." Retail followed the people, leading first to strip malls along the arterial roads into the city, then full-scale shopping malls.

The exclusions were also changing. On the one hand, the social meaning of "whiteness" had been expanded by World War II from just White Anglo-Saxon Protestants to include Jews and European Catholics. The previous restrictions on Germans and Irish were gone. While some streets still excluded Jews, the blanket restrictions were gone, and major Jewish institutions were planted outside the Highlands. On the other hand, the unspoken lure of the suburbs for white people was that it was far removed from black people, who were still restricted from following out of the ghetto.

The Louisville population continued to grow from 1930 to 1960, from 300,000 to almost 400,000. Then, in the 1970s, the population of the city declined for the first time. Nearly all new housing since the war was built in the suburbs, outside official city limits. The little cities, even small subdivisions, around the edge of Louisville started incorporating as home-rule cities to avoid annexation by the big city.

The biggest push factors sending middle-class people, black and white, out of the city, were the race riots and busing. These two events spurred massive white flight out of the city to the suburbs, and out of the public schools to private schools and to adjacent counties. The black middle class had fewer options of where to go to, but most left the core black neighborhoods in the west end, which deteriorated into a slum of concentrated poverty.

In the 1950s and 1960s, federally funded urban renewal led to the demolition of poor neighborhoods, mostly African American, on either side of downtown. The long African-American fight against segregation and disrespect created the preconditions for an explosion. When the police overreacted to a protest in May of 1968, a mostly black mob looted stores and set fires. The National Guard was called in to restore order. Thereafter, white businesses left downtown, and white residents left the west end.

In 1975 the court ordered the city and county school districts to merge and create a plan for integrated schooling. When black students were bused to the suburbs and white students were bused to the city, a mostly white mob rioted (Yater, 1979, 233). The massive resistance to school integration by white people, especially in the suburbs, would later have a happier, though more complicated, result in Louisville than it did in most cities. In the short run, though, the city public schools suffered, and the private schools were given a boost, with enrollment rising 22% in 1975 alone (K'Meyer, 2009, 268). The Catholic school system and the secular Collegiate School remained in the city, though their students increasingly came from the suburbs. Two new private schools in the far suburbs greatly expanded in the wake of school integration. Kentucky Country Day School, a merger of several older institutions, moved in 1978 to the far edge of the east end, just inside the new outer perimeter road, I265, now known as the Snyder expressway. Christian Academy of Louisville is even more directly a response to busing. Created in 1975 by a just-resigned public school administrator, it was first housed in a white church in an inner suburb. It occupied the abandoned campus of Kentucky Country Day in an east-end suburb. Later it built a large campus outside the Snyder, with branches in other outer suburbs.

The seed of the boburb

These factors, pull and push, led the Highlands to begin to deteriorate, starting with the portions closest to downtown. Large houses were cut up into apartments or torn down to build cheaper rental housing. Long-standing businesses moved to the suburbs or closed altogether. Public transportation declined. The downtown department stores, which had been convenient to the Highlands, moved to the distant malls or folded. The old people left, and the new ones coming in were poorer, more transient, less committed to the neighborhood.

This was the first real era of displacement in the Highlands. The stable middle class moved away. Some saw this as a gain. 1966 resident Ken Pyle said "it was a perfect neighborhood with a lot of working class folks. In the mid-1960s it was affordable, too" (Thomas, 2003, 255). Hippies moved in, and remained through the early 1970s. On the whole, though, many of the new residents were not as invested in the neighborhood, and even if they were, they did not have the resources to keep it up. This decline in the neighborhood in turn spurred even faster decline, as long-time residents sold out before they lost the value of their homes. Speculators and absentee landlords bought up these cheaper properties, seeking rent with little investment.

The silver lining of this decline was that the Highlands could become Louisville's first real bohemian district. In the 1970s and 1980s, only a century after their origin as the tony outer suburb of the city, even the Cherokee Triangle was beginning to be colonized by starving artists. One Highlands artist recounted,

> the first time I came to Louisville [in the 1970s], we came up here to see a show. We came down Broadway, took a right, went about three blocks, I said "this is where I want to be. There's something about Bardstown Road." Just the feeling of that. Part of that maybe was the clutter—the wires, those crazy lights that hang over the street half a block all the way down. There was just a comfortable feeling about all of that. We toured quite a bit, but that Bardstown Road was the Spot.

They were drawn to the "jangle" of Bardstown Road, and to their proximity with other artists.

On Bardstown Road, the empty storefronts became home to low-end bars, restaurants, and art spaces in which the bohemians could practice and work. An activist minister and his playwright wife opened The Roundtable Theater, which morphed into The Storefront Congregation. This was a performance space, coffeehouse, and anti-Vietnam War forum. Years later they would open

The Rudyard Kipling restaurant, the "center of the avant-garde in Louisville," with the help of their neighbor, who headed the city's Landmarks Commission (Kaukas, 1992, Scene section 5).

This tiny artist colony existed amidst a much larger population of long-time residents, some working-class people moving in for lower rents, and people just passing through. But the bohemians, for the first time in Louisville's history, had a critical mass in one place to begin to see themselves as a community. The city's leading newspaper, the *Courier-Journal*, marked the existence of "the young bohemians of Louisville" for the first time in 1976 (Winerip, 1976, 6).

Developments in adjacent neighborhoods had an impact on the Highlands – both directly and as a cautionary tale. In the 1960s, the University of Louisville expanded, and significant numbers of students began to live in Old Louisville. The big houses on leafy streets of the city's first mansion district were being cut up into apartments. The transient population had few informed "eyes on the street" looking out for the neighborhood. The nearby federal projects were declining into drug and crime centers, which spread petty crime into Old Louisville. This, in turn, accelerated the decline of the neighborhood. There was a brief moment in the 1960s when Old Louisville might have been the core of a bohemia. In 1968 a British journalist, who lived there, wrote

My neighbors are a bunch of swingers – everyone plays guitars, signs petitions and eats Mexican food. ... Our friends, the outsiders who come to parties on Belgravia Court, admire our stand for liberalism. They say the same about the Court itself: they would love to live there, too, but they are family people. "And the schools, my dear." And those things – heard of and imagined – which go on in Central Park. With the smugness of downtown dwellers, we tell them that we couldn't possibly live anywhere else.

This reflection was offered as she was returning to Britain to write for *The Guardian* (Eades, 1968, 172). Alas, after the nearby race riots later that year, Old Louisville suffered a greater decline than neighborhoods further out.

By the same token, the expansion of the University of Louisville also created a new population of students who could hang out on Bardstown Road, three miles distant. The expanded University of Louisville faculty could afford to buy in the now-cheaper Highlands. A long-time English professor at the university, looking back in the early 1980s on a grittier previous era, noted:

Parts of Bardstown Road and the lower Highlands probably offer the most visible evidence of today's bohemia: Waterholes that range from trendy to tawdry; second-hand furniture shops; stores that cater to young musicians. Housing is diverse – big old houses here and there; inexpensive apartments in the shabbier streets.

(William Axton, quoted in Woolsey, 1983, 146)

Urban university professors were exactly the constituency that appreciated the virtues of a walkable, leafy, mixed-use neighborhood, which had the added benefit of being near their job. These urban knowledge workers were strategically important both in patronizing bohemia, and in laying the foundation for boburbia.

In 1963 the Cherokee Triangle historic preservation association formed, when they saw the failure to save Old Louisville. This is about when the term "Cherokee Triangle" starts to be used. *This is the nucleus of the Highlands as a boburb.* The neighborhood had existed for a century, but the effort to defend it led the remaining long-term residents to develop a new self-consciousness and new sense of purpose. The Highlands began to emerge as a *project*, which afforded residents with new tools to create a distinctive place (Weston, 2018). This new purpose, in turn, drew a new kind of resident, drawn to both the bohemian vibrancy and to the bones of a dense, walkable residential community.

Because of the importance of this seminal moment in reforming the Highlands as a boburb, let's follow this story in a bit more detail.

"It was the yuppie place to live," John Anderson remembers.

We paid a good price for our property and put a lot of money into the house and yard, and our neighbors were doing the same. We felt we ought to have the support of the Planning and Zoning Commission in that effort. The first stages of revival were beginning to happen in a lot of areas adjacent to downtown.

(Thomas, 2003, 219)

Anderson, who worked for the Planning and Zoning Commission and lived in the neighborhood, became the chief advocate for the association. Their letter to the planning commission said:

the Cherokee Triangle is inhabited by people who live there by choice – not economic necessity. These are the people who have given the neighborhood its distinctive character, which we believe to be well worth preserving. But, if little by little, non-residential uses appear along Cherokee Road, we shall surely have created another Third Street [Old Louisville]. Another convenient and pleasant neighborhood will have permanently disappeared, for this kind of change is irreversible.

(Thomas, 2003, 220)

The mobilizing factor for the whole neighborhood came when a prominent Victorian house was demolished and replaced with an undistinguished apartment building. When another such structure was threatened, the neighborhood association protested, and began working on legal protection, by getting historic neighborhood sections declared "landmarks." In 1972 a city-wide preservation alliance was created. It consisted of 28 neighborhood associations, some government agencies, professional societies, and civic, educational, and historical groups. Architect and Cherokee Triangle resident John Cullinane was executive director. The Cherokee Triangle Association was one of the founding members (Thomas, 2003, 226). One argument for landmark status, offered by University of Louisville law professor Nathan Lord, begins to articulate the boburb ideal: "In an era of recession, inflation, pollution, and transportation problems, we have a workable alternative to complete suburbanization" (Thomas, 2003, 228).

In 1975, the Cherokee Triangle itself was granted landmark designation. Stabilizing the residential heart of the Highlands helped spur economic development on Bardstown Road. The empty storefronts and junk stores started to be replaced by better independent stores and social gathering places. Trendy restaurants came to Bardstown Road, starting with the Bristol Bar and Grille in 1977. The Storefront Congregation was a coffeehouse and anti-war organizing node. In 1978 Carmichael's Bookstore opened on Bardstown Road, and in 1983 it moved to the corner of Longest Avenue; in 1994 the first Heine Brothers coffeehouse would be added to the bookstore. In 1975 musician Harry Bickel bought a house on Deering Court, and let so many other musicians live there that it became known as "the Bluegrass Hotel." Willow Park, the tip of Cherokee Park in the middle of the Triangle, became home to an annual art fair in 1972, and to summer concerts starting in 1982.

Over time, the Highlands became so successful as the cultural center of Louisville that more bourgeois bohemians moved there – which in turn gentrified the living costs beyond the reach of the next generation of artists and bohemians.

The fifth fifty years: the Highlands reborn as a boburb, 1980–2030

The Highlands today are in their golden age as a boburb. They made the successful transition from a seedy and struggling bohemia to hold on to their role as a cultural center for the city, while reclaiming their old economic stability. Whether they stay poised between bohemia and suburbia, or evolve into something else, will depend on internal cultural forces and external economic forces. Moreover, the Highlands neighborhood does not exist alone, but in an ecology of Louisville neighborhoods, each of which competes with the Highlands to become the new boburb, or to displace the boburban ideal.

The changing face of one commercial site on Bardstown Road gives a shorthand version of the cultural changes in the Highlands from bohemia to boburbia. 1047 has been a bar and music venue for decades. It is just a few blocks from Highland Coffee, in what is now Restaurant Row. In the 1970s it was a boogie bar with the hippie moniker Funktion Junktion. In 1981 Doyle and Mary Guhy bought it and gave it the Irish name Tewligan's, like other Irish-named pubs in the Original Highlands. They hosted a Poets' Night, which turned into Artists' Night when the singer-songwriters displaced the poets (Quinlan, 1985). Tewligan's had its greatest fame as "ground zero for the original music scene" in Louisville in the late 1980s and early 1990s, hosting local and national punk bands (Puckett, 2017). Tewligan's closed in 1996, and was succeeded by Cahoots, a bar and pool hall. Cahoots was forced to close due to the customers' regular violence and drug use. In 2017 the space reopened as an Indian-inflected restaurant and dance club, Nirvana. Kashmira Singh, who also owns Kashmir Indian Restaurant a few doors down Bardstown Road, said "It will be upscale, casual atmosphere targeting millennials" (Bowling, 2017). Cosmopolitan boburbanization took another step forward.

The tipping point toward full boburbia might have come about 1990. One measure of the change at this time is the story in *Louisville* magazine in July 1988, which called the Highlands "Louisville's best-kept secret." The story praised the neighborhood's well-kept older homes, sidewalks lined with mature trees, and "a population that has an appreciation for the area's urban tinge as well as its greenery." Thirty years later the same magazine lifted this story up as a flashback to a time when the Highlands was a secret, to show how much things had changed (Brewer, 2018). By 1992, the bourgeois businesses on Bardstown Road had found that bohemianism was a strong selling point. Their ad in the *Courier-Journal* claimed a vital cultural role for "the Road" within the whole city:

> Stretching from the heart of our city into the suburban foothills of the county there is a vibrant, pulsating lifeline called Bardstown Road. It is on and along this teeming byway that "Business as Unusual" daily plays out its story.
>
> (Lee, 1992)

That central type of bobo venue, the coffeehouse-in-a-bookstore, soon followed. In 1992, Twice-Told Tales, a used bookshop, added a coffeehouse, drawing on bohemian models. They had bright reading-and-sipping rooms, dark poetry-and-music performance rooms, framed "hip" magazines on the wall, newspapers on the counter, and National Public Radio on the speakers (Puckett, 1992). The Twice-Told coffeehouse was an "instant hit" – "This is the place where you can find Express-clad matrons rubbing padded shoulders with skateboard boys in baggy britches. Old hippies, new hippies, poetic hipsters and hip retro-nerds are among the regulars." Two years later, Carmichaels, which sold new books, added Heine Brothers Coffee in their back room. Gary Heine articulated a strong boburban theme that the local is more authentic:

> I think people have a sense that whatever you can get that's processed closer to home is going to taste fresher and therefore better. ... I think there's also a kind of community feeling about buying from your

neighborhood coffee roaster or beer maker that people really want to have. Something about being part of the community.

(Lundy, 1993)

In the boburb, community is something you make locally.

When the Highlands promoted the slogan "Keep Louisville Weird," it crowned its position as the city's premier boburb. The slogan was borrowed in 2005 from the "Keep Austin Weird" movement in Texas (Long, 2010). John Timmons, the owner of Ear X-tacy, along with a few other local business owners, created the Louisville Independent Business Alliance (LIBA) around the slogan. LIBA are the chief promoters of both "buy local" campaigns in the city, and pro-quirkiness efforts in the Highlands specifically (Shafer, 2015).

One measure of the strength of the Highlands as a neighborhood today is that it held its own in the Great Recession that began in 2008. There were almost no foreclosures in the Cherokee Triangle, and houses retained their value while most other neighborhoods lost ground (Besel, 2013).

In the 21st century, the boburbs have become the center of gay and lesbian institutions in Louisville. At the end of the 20th century, the gay and lesbian bars were clustered in the declining commercial area around Market Street, the area that would later be revived as NuLu. In that era, queer institutions still lived a semi-underground existence removed from residential areas. The short-lived Louisville Gay Liberation Front was founded in 1970 in Old Louisville, organized in downtown bars, but petered out after police raided a "Gay Lib" house in the Highlands in late 1971 (Fosl, 2012). By the 2010s, though, in a new era of acceptance for a variety of sexualities, nearly all the old gay and lesbian bars were closed. In their place, a new generation of gay-affiliated bars, restaurants, and coffeehouses had opened on Bardstown Road in the heart of the Highlands (Lauer, 2017). The Fairness Campaign, which was founded in 1991 to fight sexual orientation discrimination, is head-quartered in Louisville's other boburb, on Frankfort Avenue.

The elephant in the room: race and residence

Race always matters in America, but since the great successes of the Civil Rights Movement it has been declining in significance in determining the life chances of everyone, especially African Americans (Wilson, 1978). Race is giving way to class as the main shaper of who is included and excluded in social groups in Louisville. The Highlands, in particular, values diversity, and wants to include a wider variety of races and ethnic groups of people who share the class culture of the neighborhood. In all the middle-class neighbor-hoods in the region, in fact, I found that families of all races who can afford the freight are welcome. The choice of neighborhood lies mostly with the households and their bank accounts.

The bigger fight about race and class comes in the choice of schools (this section draws extensively on Johnson, 2013). This is because the choice of schools is *not* entirely in the hands of the family. This is especially a source of anxiety for people who like to control all the conditions of their lives and control who their children come in contact with. This takes us into the weeds of Louisville's unusual school assignment system.

Louisville's schools were racially segregated from the beginning, almost entirely to separate African-American students from everyone else. In 1904 Kentucky's notorious Day Law made integrated education illegal in any school, public or private. The US Supreme Court invalidated this law and other racial segregation laws in 1954 in the landmark *Brown v Board of Education* decision. Thereafter, Louisville, like other segregated cities, tried various plans to desegregate its schools. In 1975 a judge ruled all these efforts inadequate, since "white flight" meant white people were evacuating the city schools for the separate Jefferson County school district. The judge ordered a merger of the two districts, and a more powerful plan to create racial balance in each public school in the new district. As previously noted, this plan met with massive resistance, especially by petite bourgeois white people, and the creation of white private schools far from the city. The fight, and the white (and black middle-class) flight continued for another three decades.

In 2007, in response to a lawsuit initiated by Louisville parents, the Supreme Court made a crucial change in the *Brown* legacy. In *Meredith v Jefferson County*, they ruled that desegregation plans based on race alone were unconstitutional. This obliged the school district, in Louisville and many other cities, to come up with a new plan. Instead, the county was divided into six clusters, each with a rich and a poor reside area. All the kids in each reside area are classified as the same category, regardless of their individual race or class. Every household had a "reside" school nearby, but each cluster also had schools in the other side of the cluster that parents could choose. Children applied to get into a school, with weight given to applicants from the other side who would help balance the school. Families were not guaranteed that their child would be admitted to their reside school, even if it were closer, depending on the balance of all the schools.

Magnet schools were excluded from the cluster system, most especially Manual, Male, and James Graham Brown high schools. They could choose students from anywhere in the district, based on a weighted lottery. One of the weights that helped a child get selected for a magnet high school was if they had attended a magnet elementary school. Crucially, the magnet elementary schools were in the poorer and blacker parts of town. Thus, families had some choice of where their children went to school, but were also subject to institutional rules that could limit those choices, year by year and child by child. In general this system has worked fairly well to keep Jefferson County schools more integrated, and less strife-ridden, than many districts that are under integration orders (Semuels, 2015).

The most careful study of how, exactly, parents make these decisions found a difference between liberals and conservatives, which generally, though not exactly, mapped on to the boburb/suburb divide. None of the white parents, and few of the black parents, were comfortable talking about race as relevant to their decisions. The more liberal white parents, and most of the black parents, though, wanted to send their children to the public schools, and wanted a diverse and integrated education for their children. The knowledge class, in particular, depended on the magnet high schools for their children, both because of their lower cost than private schools and their ideological claims to be more civic-minded.

Conservative and suburban parents, by contrast, were more likely to be ready to bolt the public system if it meant their children would be assigned schools in poor and minority areas that they regarded as dangerous. To such parents, the private schools, or moving across the county line, were often the first choice, or the default response to what they regarded as an adverse assignment by the school district. These parents never described their choices in terms of race, but rather of the logistical inconvenience of having their small children in schools so far away from where the parents lived or worked, or the low achievement scores often found in the schools in poor and high minority parts of town.

Two east end mothers, one white, one black, had children assigned to a west end elementary school. Both professed a desire that their children have an integrated education. The black mother decided that, though she worried about her child having a long commute, the school was near enough to the mother's downtown job that they could make it work. The white mother, though she professed "liberal hippie values," decided the logistics of transportation were reason enough to opt out of the public system. Other suburban parents simply refused to consider west end schools for their children, and did not even list them in their preference list for school assignment (Johnson, 2013, 135ff).

The Highlands parents, by contrast, were more likely to support the public schools and the district's integration system. The Highlands parents organized to save Bloom Elementary, adjacent to Highland Coffee, and treasured it as a community institution. Two white liberal mothers voted with their feet when they moved to Germantown with small children. Not surprisingly, they welcomed their children's assignment to west end elementary schools with special programs, which would also give their children a better shot at the magnet high schools down the road (Johnson, 2013, 159, 190).

A walk to all the coffeehouses

We began our stroll of the boulevards of the boburb with a walk to Highland Coffee. This is but one of many coffeehouses of the Highlands, each serving a somewhat different micro-neighborhood and coffeehouse constituency. There are four miles of Bardstown Road from the Original Highlands to the Watterson, from the intersection with Broadway to the on-ramp of the interstate. In those four miles are eight independent coffeehouses in 2019. Each is distinct, even through three are from the same local chain, Heine Brothers.

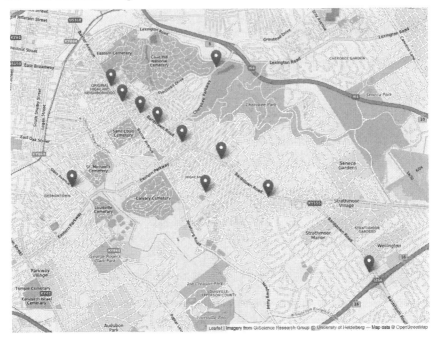

Figure 3.2 The coffeehouses of the Highlands

Quills Coffee House is in the heart of the Original Highlands. The Quills story serves as a perfect example of how bohemian culture is normalized into a bourgeois work ethic – the bobo formula. Nathan Quillo was a social worker who met clients in coffeehouses. He developed a passion for coffee, which led to a tiny store he opened with his brother. In 2007 they opened in Germantown, the cheaper working-class neighborhood that is turning bohemian. Two years later they moved to the Original Highlands, into an existing commercial strip of bars, clothing stores, and music venues. They were serious about coffee origins, roast, and preparations. They briefly hosted a vinyl record store in the back, catering to hipsters who thought retro music technology made a more authentic sound. Within a decade they had added stores at the University of Louisville, another across the river in New Albany, IN, and a third far flung outpost in Indianapolis. They opened a store in the new hipster project of NuLu, on the edge of downtown, and in St. Matthews, on the edge of suburbia. They had added a roaster, with retail and wholesale sales of roasted coffee, along with tee shirts with a coffee-drinking porcupine urging you to "Stay Sharp" (Quills Coffee, 2018).

In 2007, the same year that bohemian Quills began, corporate coffee behemoth Starbucks opened a store right at the base of Bardstown Road. The location is revealing, and anomalous in the Highlands. The building it is located in describes itself as the site of Louisville's first suburban shopping center, built in 1936. Completely rebuilt in 2000, it offers a chain store pharmacy, fast-food shop, and coffee, with the suburban draw of parking spaces on the property (Dahlem Properties, 2018).

The genius of the "creative class" comes out in combinations of different genres. In 2014 the Gralehaus opened in a Victorian mansion, combining a coffeehouse, craft beer tavern, and bed and breakfast, adjacent to the Holy Grale pub, built in an old church. This lets them offer new combinations for adventuresome tastes: "soft serve affogato [coffee over ice cream], honey bee pollen cortado, and sorghum latte," as well as "a special 'coffee-rocket' draft line (beer infused with coffee beans)" (Gralehaus, 2018).

Highland Coffee and Bakery is a fixture of the Tyler Park neighborhood within the Highlands. Opened in 2000, it bills itself as "Louisville's Finest Coffeehouse." The four-foot wide coffee cup over the front door is a land-mark. They proudly advertise that their bakery is "mostly VEGAN!" (Highland Coffee, 2018).

A mile into our journey we come to the original Heine Brothers store on Longest Avenue. Heine Brothers was begun in the early 1990s by two men with a passion for specialty coffee and a hand cart in the mall. In 1994 they opened their first store. In true boburban fashion, it connected internally with a local independent bookstore, Carmichaels. Heine Brothers beat Starbucks to Louisville, and established themselves as the leading local chain. Today they have their own roastery, more than a dozen stores all in greater Louisville, and give extensive support to all aspects of independent business culture in the city. This is the first of three Heine Brothers stores in the Highlands, each of which has a different look and feel. In 2018 they bought out and demolished the Dunkin Donuts store across the street, moving the tiny original shop to a much larger new building with a drive through (Heine Brothers, 2018).

Day's Espresso and Coffee serves the southern edge of the Tyler Park neighborhood. Begun the same year as the first Heine Brothers store, it maintains a low-key social media presence, consisting only of pictures. They claim to have the best espresso in town. They have a very traditional coffee menu. They partner with the local cookie shop next door, and the local bagel shop across town. Day's is also known as the most gay-friendly coffeehouse, though they do not want that to be their main identity (Everson, 2009).

Safai Coffee describes itself as "in the heart of the Highlands." It is on the boundary between the Bonnycastle and Deer Park neighborhoods, and has a sizable group of walk-in regulars. Its self-presentation is for coffee purists. The owner contends that "organic Microlot grade coffees are the most premium coffee beans on earth." He samples many growers' offerings each year, then purchases one farmer's entire yield for the year. Though they provide many machine-pressed espresso drinks, this coffeehouse favors the pour-over method, which they have on prominent display, as "the best brewing method to showcase each varietal's distinct characteristics" (Safai Coffee, 2018).

Two and a half miles down our route we come to Heine Brothers at Dou-glass Loop. This neighborhood got its name from the turn-around for streetcars to head back toward downtown. The coffeehouse is located in the small block circled by that loop. It is the most spacious Heine Brothers, with several big tables, couches around a stove, and a welcome to all, including the well-known homeless man who spends most of each day there.

Finally, at the end of the Outer Highlands, just before the on-ramp to Interstate 264 that circumscribes Louisville, is Heine Brothers – Gardiner Lane. This store is at the interstate end of a strip mall. A large part of its business comes from a drive through for commuters coming in to the city. Of the several Heine Brothers shops on this route, this one feels the most like a Starbucks.

Appropriately, on the other side of the interstate is an actual Starbucks, which is designed for easy access by suburban commuters. But that territory lies beyond the boburb, and begins the true suburb.

References

Karl Besel, 2013. "Louisville's Historic Belles: Cherokee Triangle and Old Louisville," in Karl Besel and Viviana Andreescu, eds. *Back to the Future: New Urbanism and the Rise of Neotraditionalism in Urban Planning*. Lanham: University Press of America, 49–56.

Stephanie Bower, 2016. *Kentucky Countryside in Transition: A Streetcar Suburb and the Origins of Middle-Class Louisville, 1850–1910*. Knoxville: University of Tennessee Press.

Caitlin Bowling, 2017. "New Bar and Restaurant Concept to Open in Cahoots." *Insider Louisville*, January 25.

Megan Brewer, 2018. "Flashback: July 1988." *Louisville Magazine*, 69, 7, July 21.

City Planning Commission [Louisville], 1929. *A Major Street Plan for Louisville, Kentucky*.

Barbara Conkin, 2003. *Why Are the Highlands High? The Geology Beneath the Landscapes of Jefferson County*. Louisville: Hycliffe Publishers.

Dahlem Properties, 2018. *1000 Baxter Avenue website*. https://www.dahlem.com/properties/1000-baxter-avenue-center/.

Christine Eades, 1968. "Togetherness in Belgravia Court." *The Guardian*, May 12, 172.

Carol Ely, 2003. *Jewish Louisville: Portrait of a Community*. Louisville: Jewish Community Federation of Louisville, Foundation for Planned Giving.

Judith Hart English, 1972. "Louisville's Nineteenth Century Suburban Growth: Parkland, Crescent Hill, Cherokee Triangle, Beechmont and Highland Park." Unpublished MA thesis, Division of Humanities, University of Louisville.

Zav Everson, 2009. "Ditch the Stereotype: Louisville is Gay-Friendly." *Louisville Magazine*, 60, 5, May 15.

Catherine Fosl, 2012. "'It Could Be Dangerous!' Gay Liberation and Gay Marriage In Louisville, Kentucky in 1970." *Ohio Valley History*, 12, 1 (Spring): 45–64.

Gralehaus website, 2018. http://gralehaus.com/.

A. Gwynn Henderson and David Pollack, 2012. "Kentucky," in *Native America: A State-by-State Historical Encyclopedia*, Volume 1, edited by Daniel S. Murphree. Santa Barbara: Greenwood Press.

Heine Brothers Coffee website, 2018. https://heinebroscoffee.com/heine-brothers-relocating-cafe-at-bardstown-rd-and-longest-ave/.

Highland Coffee website, 2018. https://twitter.com/highlandcoffee?lang=en.

J. Blaine Hudson, 2011. "'Upon This Rock' – The Free African American Community of Antebellum Louisville, Kentucky." *Register of the Kentucky Historical Society*, 109, 3 & 4 (Summer/Autumn): 295–326.

Kenneth Jackson, 1985. *Crabgrass Frontier: The Suburbanization of the United States.* New York: Oxford University Press.

Rebecca Page Johnson, 2013. "Desegregation in a 'Color-Blind' Era: Parents Navigating School Assignment and Choice in Louisville, KY." Cultural Foundations of Education – Dissertations. Paper 56. PhD, School of Education, Syracuse University.

Dick Kaukas, 1992. "The Rud. Bar? Restaurant? Arts Center? One Thing Is Sure: There's Nothing Else Remotely Like It." *Courier-Journal*, February 1, Scene section, 5.

Tracy K'Meyer, 2009. *Civil Rights in the Gateway to the South: Louisville, Kentucky 1945–1980.* Lexington: University Press of Kentucky.

Landon Lauer, 2017. "An Analysis of Gentrification's Effects on LGBTQ+ Populations in Louisville, Kentucky." University of Louisville, College of Arts & Sciences Senior Honors Theses, Paper 128.

Stephen Lee, 1992. "For Great Neighborhoods, Restaurants, Shops, and More Bardstown Is the 'Road' to Travel." Bardstown Road Aglow Advertising Section, *Courier-Journal*, December 2.

Joshua Long, 2010. *Weird City: Sense of Place and Creative Resistance in Austin, Texas.* Austin: University of Texas Press.

Ronni Lundy, 1993. "Coffee Is Hot." *Courier-Journal Scene*, January 23, 15.

Anne Marshall, 2010. *Creating a Confederate Kentucky: The Lost Cause and Civil War Memory in a Border State.* Chapel Hill: University of North Carolina Press.

Jeffrey Lee Puckett, 1992. "Twice Told Coffeehouse Is More Than a Java Stop." *Courier-Journal Scene*, October 3, 9.

Jeffrey Lee Puckett, 2017. "From Punk Rock to Puri: A Piece of Louisville's Music Scene Lives On With Nirvana." *Courier-Journal*, June 23.

Quills Coffee website, 2018. https://quillscoffee.com/pages/about-us.

Michael Quinlan, 1985. "'Artists' Night' at Tewligan's Lets Performers Exhibit Talents from Chanting to Playing Music." *Courier-Journal*, April 5, C1.

John Rogers, 1955. *The Story of Louisville's Neighborhoods.* Collected stories from the *Louisville Times* of May 1955, published by the *Courier-Journal* and the *Louisville Times*.

Mary Sachs, 1997. "Castlewood: A Late 19th–Early 20th Century Suburb of Louisville, Kentucky. The Development of a Neighborhood from Geographic and Cultural Perspectives." Masters Thesis, Interdisciplinary Studies, University of Louisville.

Safai Coffee website, 2018. https://safaicoffee.com/.

Alana Semuels, 2015. "The City That Believed in Desegregation." *The Atlantic Monthly*, March 27.

Sheldon Shafer, 2015. "Group to Celebrate Keeping Louisville Weird." *Courier-Journal*, February 9.

Samuel W. Thomas, 2003. *Cherokee Triangle: A History of the Heart of the Highlands.* Louisville: Cherokee Triangle Association.

C. Robert Ullrich and Victoria A. Ullrich, eds., 2015. *Germans in Louisville: A History.* Charleston: History Press.

Sam Bass Warner, Jr., 1962. *Streetcar Suburbs: The Process of Growth in Boston, 1870–1900.* Cambridge: Harvard University Press and the MIT Press.

William Weston, 2018. "Are Neighbourhoods Real?" *Journal of Critical Realism*, February, 34–45.

William Julius Wilson, 1978. *The Declining Significance of Race: Blacks and Changing American Institutions*. Chicago: University of Chicago Press.

Mike Winerip, 1976. "Sweet Shop: Anybody Can Bake A Cake, But ..." *Courier-Journal*, July 24, 6.

F.W. Woolsey, 1983. "Axton's Axiom." *Courier-Journal*, August 14, 146.

George Yater, 1979. *Two Hundred Years at the Falls of the Ohio: A History of Louisville and Jefferson County*. Louisville: The Heritage Corporation of Louisville and Jefferson County.

George Yater, 1984. *Flappers, Prohibition, and All That Jazz: Louisville Remembers the Twenties*. Exhibition catalog for the Museum of History and Science. Louisville: Museum of History and Science.

4 Why do you live in the Highlands?

The core question I asked all the people I interviewed was "Why did you choose to live where you live?"

I believe that our choice of where we would like to live reflects many factors, some shallow, some deep. Material reasons – how much does it cost for the space? how long is the commute? what would the resale value be like? – are relevant. More cultural reasons also figure, such as the quality of the schools, the beauty of the house, the level of community interaction. And deep matters of worldview shape our ideal neighborhood, as well. These deep matters are not irrational emotional reactions. Rather, they are rational judgments embedded in quick emotional responses, which are so immediate that we can't really articulate the reasons behind them (Haidt, 2012).

When I started this study, I interviewed several real estate agents. I thought they might be taste-makers, steering people to one kind of neighborhood or another. I was quickly corrected – according to the real estate agents, people already know what *kind* of neighborhood they want to live in; they want the real estate agent to help them find an affordable property that matches their choice. For the most part, this was true of my interviews, as well. When I asked people the survey question noted above – "Imagine you are moving to a new city ..." – nearly everyone could answer this question clearly and quickly. This was true even if where they currently lived was not the same as where they would like to live.

For most white college graduates, their expectations and the real estate agents' assumptions matched, so that neither party was obliged to be conscious of what those expectations were. Both stayed away from the poorest, most dangerous areas, whether those neighborhoods were predominantly white, black, or other (mostly immigrants). They instead steered each other to the boburbs if they were young and looking for an exciting social life, or into the eastern suburbs if they were thinking about children or were just more cautious about "the city." It is part of middle-class white privilege that the path of least resistance leads to white, middle-class neighborhoods, even if you are not conscious that that is what you are looking for (Johnson, 2005).

The non-white people I interviewed were more conscious of the whiteness of the city's main middle-class regions. However, *none* of the non-white people I interviewed reported being steered toward predominantly non-white neighborhoods, nor away from the white suburbs. The old-fashioned kind of redlining and racial steering is, for college graduates, at least, largely a thing of the past. Rather, the consciousness of race came from the home-buyers themselves. Non-white people were more likely to ask for a more diverse neighborhood, especially if they were liberals. The story of choosing where to live, and what kind of welcome they would get, might have been different in a more working-class world, but for these college graduates, middle-class civility was the norm.

All the middle-class African-American families I talked to lived in predominantly white neighborhoods. By their account, this was because they definitely wanted to live in predominantly middle-class neighborhoods, all of which in Louisville are predominantly white. However, several told me they did not want to be the *only* black family in the neighborhood. Working with real estate agents and a grapevine from church, fraternities and sororities, and work mates, they identified specific streets and subdivisions that were more integrated than average.

An organization of some of the richest African-American families shows a distribution like other upper-middle-class people, with a twist. Ten percent lived in the Highlands or Old Louisville. Two-thirds lived in the east end, with almost half of them outside the Snyder. The twist is that a fifth still lived in the west end, where their families provided professional services to the black community. They tended to live in the far west end, near Shawnee Park, in the residue of the old mansion neighborhood from the streetcar era (confidential personal communication).

The other kind of people who eschewed the very white, very middle-class or upper-middle suburbs were white liberals. Some moved to the suburbs first, mostly for family reasons, then gradually came to feel stifled or guilty that everyone looked like them. Others said from the outset that they did not want to be in the "regular" suburbs. Several reported that they had to fight to get the real estate agent to believe that they really did not want to see anything in the white east end – sometimes resulting in a change of agents. These are the kind of people most likely to raise their families in the boburb.

The people who chose to live in the Highlands almost always did so knowing the reputation of the neighborhood. This was also true of some who wished they could live there, but couldn't afford it or were obliged to live elsewhere. The reputation of the Highlands was also important to some people who deliberately chose against it, either for bohemia or, more often, for suburbia. Several people told me that they lived in the suburbs when they first moved to Louisville because they thought that was what they wanted, or thought that was just what one (in their class) did, or were steered there by real estate agents who assumed that was what they wanted, or, in a few cases, by people who just made a mistake. There was an asymmetry between choosing the boburb – the Highlands or Crescent Hill – versus choosing a suburb. The boburbs were specific neighborhoods, with distinctive special features. The suburbs were a *type* of neighborhood, which could be met by any one of a number of subdivisions. Unless you were trying to live near relatives or near a job, few people were committed to one specific suburb, in the way that other people were committed to a specific boburb.

A suburb is a kind of space. A boburb is a specific place.

Who chooses the Highlands?

People choose the Highlands because it is a distinctive neighborhood. From interviews with residents the attractive features that stand out the most were the many interesting places to go and things to do that were in walking distance. They liked the old and different houses, the big trees, the culture, the liberal politics, the sheer density of social interaction. Some of the challenges they noted were partly those of city living – noise, dirt, traffic, parking, caution about crime. Other challenges came from succeeding too well: so many people wanted to live there that younger and poorer people were being priced out, and the mom-and-pop local stores were being driven out by the chain stores. Those who had arrived there in the 1970s thought of themselves as pioneers; they had come in at the bottom and helped rebuild the neighborhood. They only worried that the Highlands were becoming too bourgeois. Those who had come more recently were glad to find a place for themselves in the Highlands. They only worried that others like themselves had to go elsewhere. And all of those who liked the Highlands praised the diversity of kinds of people one could find there.

Consider these voices of contented residents, representing a spectrum of ages, occupations, and political views.

First, the people you would expect to love it the most – liberal artists:

The first time I came to Louisville [in 1978], we came up here to see a show. We came down Broadway, took a right, went about three blocks, I said "this is where I want to be. There's something about Bardstown Road." Just the feeling of that. Part of that maybe was the clutter—the wires, those crazy lights that hang over the street half a block all the way down. There was just a comfortable feeling about all of that. We toured quite a bit, but that Bardstown Road was the Spot.

This was the person we met before, who was immediately drawn to the "jangle" of the Highlands. This is the "keep it weird" spirit. Visual artists expressed most strongly a reaction that many kinds of bourbanites shared. They like the cosmopolitan mix, variety, diversity.

People who had lived in more mixed places were drawn to the Highlands as the closest equivalent. One couple in a cross-national marriage, moving from the West Coast, initially were steered to the suburbs for class reasons. However, they later moved to the Highlands:

There's a kind of eccentricity, clutter, visually, both present in the architecture but also in the people. The chance to meet my neighbors and be connected to them, the fact that this is a porch community just creates a contact point. Coming from San Francisco, I miss the languages, the different skin tones, people coming from all over the world. We were hungry for connection.

Another artist put the appeal diplomatically. "When I say that we wanted to be connected to vibrancy, the suburbs are too clean. There's something about the density of the city that is very attractive to people who see the world in a certain way." Another successful artist was more emphatic: "I like that you see that diversity walking down the street. I don't change out of my painting clothes" to go to the store. She thought the norm in the Highlands was a willingness to engage that you don't get in the suburbs, no matter what race or gender or even if one's gender could be identified. She went so far as to say that for an artist to live in the suburbs represented too much of a compromise, was too far from the central mix that art should come from. She rejected the "Keep Louisville Weird" slogan, because she didn't think the city was weird *enough*.

Those were established older people. The appeal to young hipster artists is quite similar, with a few twists. They all praised the walkability of their neighborhood. They are not just walking in order to walk, but "having a destination and stops along the way is appealing." Moreover, being in walking distance of their favorite bars meant they could walk home, not drive, after an evening of drinking. They liked being proud of where they live, that "people want to come to your neighborhood and hang out." The 20-something artistic types in the boburb all live in a public way – "not just in your house." One person compared this public way of living in the city to how these same people lived on campus when they were in college. In a boburb, it is sufficient to live in a small residence, because sharing public facilities is a virtue and a blessing to sociability.

Old knowledge-class types, not artists but professors and scholars – bobos before the letter – also start their list of positive boburb features with the mixed use and diversity of the neighborhood. Some had been there since the more bohemian period in the 1970s. They praised walking, too, but they listed walking to the coffeehouse or the library, past interesting old houses, and under tall trees, before they mentioned bars. They praised the public transportation, the large and maintained parks, the decent-to-excellent schools. Louisville, unlike most Southern cities, has a large and active Catholic community, and many people who might have moved to the suburbs stayed in the Highlands because of the still-vibrant parish institutions. The Protestants and Jews also liked the fact that they could walk to their churches and temples.

Everyone in the Highlands praised its diversity – and immediately noted that it was not very racially diverse yet. One of the most thoughtful students of the neighborhood said "we see a breadth of different people as interesting, rather than something to insulate yourself from." In the 1970s his kids played in Tyler Park with "rougher" kids because "we were more willing to accept that as a fact of life than you would find in the suburbs." They were willing to accept that roughness, despite some apprehension, because "we are more liberal, we struggle with being more tolerant, we perceive ourselves as being more tolerant." They saw their tolerance as a reaction to the open racism they grew up with in the Louisville of the 1940s and 1950s.

Thinking about what could undermine the Highlands, he noted rising housing costs. If housing prices start at $475,000, you are "not talking about schoolteachers." Older Highlanders, who came in their 30s, and are now in their 60s, are anxious that "the funkiness" has been lost; "the little shops that never made much money but sold tie-dye." When the HopCats, Steel City Popsicles, and the other chain stores come in, "where do the skateboarders go? Where do the kids with the green hair go?"

The Highlands have the most active neighborhood associations in the city (Hur and Bollinger, 2015). Several hundred people regularly participate in them, organizing events, dealing with local issues, and making connections that are useful for other purposes. While these active volunteers are a small proportion of the thousands who live in the Highlands, they dwarf the level of involvement found in the suburban homeowners associations, where getting *anyone* to serve is often a task akin to pulling teeth.

A mainline-denomination minister who feels more "culturally comfortable" in the Highlands than anywhere else in Kentucky put the boburb's embrace of diversity in more elevated terms. The Highlands is best for "people who believe there is a blessing in the Other, especially when the interaction is not entirely predictable. I prefer serendipity to my interactions, as opposed to managing my life to the hilt." Though they had had a few minor burglaries, she was not scared to live or raise her kids in the Highlands. While she allowed that you do see "weirdos" at bus stops in the cities who you would not see in the suburbs, she asserted that "I want people like that in my vicinity and my children's vicinity, because I don't want them to think the world is only full of clean people."

A school teacher married to an arts manager had lived in the suburbs when their children were born, but moved to the Highlands because she thought it was better for the kids. The children went to elementary school in the projects, and still had friends from there. They learned to walk places, navigate public transportation ("a life skill"), stay in touch by cell phone, and appreciate "proximity of people not of the same group." The teacher liked the political liberalism of the Highlands, as well as the cultural cosmopolitanism. Her husband's job was "suspect" in the suburbs, but accepted and better understood in the boburb. She concluded "I don't think we would have been reading the same things" if they had stayed in the suburbs. She offered a novel account of why liberals like to walk: it is part of their commitment to recycling.

A social worker active in the neighborhood association was happy to renovate an old Highlands house in the 1970s, when others were fleeing for the suburbs. She was drawn to the high level of political activity – the highest in the state. She also likes that "the Highlands also has a highly educated population and is rather eccentric." Another neighborhood association stalwart said she would never move back to the suburbs because they are too far from the arts, cultural, neighborhood, and political scene.

Neighborhood association officers are among the best examples of the boburb "ideal type." One man grew up in the suburbs and hated it. Studying political science at the University of Louisville led to a career with city government. He was drawn to the neighborhood association from a sense of civic responsibility, shared by "a lot of old hippies." The Highlands, he allowed, "is an area you move to if you like being around people and like being around your neighbors." In the suburbs, he asserted, you only meet your neighbors if you invite them in. In the Highlands, by contrast, it is hard getting around the grocery store, because every aisle has people talking to one another. The coffeehouses and churches have meeting rooms, the sidewalks are full of walkers and bikers, Cherokee Park is "almost loved to death," and "just about everybody here has a dog." He acknowledged some dangers in the city, but said that he and his neighbors were cautious but not scared, not prone to "binary thinking" that any place was all safe or all dangerous. When offered the concept of a boburb to describe the Highlands today, this well-embedded long-time resident readily agreed:

> We have hippies with law degrees. This is a high rent district. It has elements of Cambridge in it. Maybe Somerville. This used to be bohemia in the '70s; not now. Now, people here go to Shakespeare in the Park, give to KET [Kentucky Educational Television], have tickets to Actors Theater. They don't stay up all night playing saxophone.

It is not only the artists and intellectuals who love the boburb.

The interesting symmetry of two single women medical professionals, of different generations, shows another appeal of the diverse community. One grew up in a conservative Christian community in the far suburbs. She almost married into a similar role, but decided she didn't want to reproduce that life. She bought a Highlands bungalow on her own, has a downtown career, and her parents worry about her living in the city.

The other woman grew up in the city, married, and had a child. However, when she was widowed, and the rest of her family had fled the city for a rural compound, she stayed. She likes the community of the Highlands. She loves the charm of old houses, which she thinks new ones can't replicate. She knows her neighbors, has coffee on their porches in the evening, shares a neighborhood Christmas party. Her child never played in the yard, because the whole neighborhood was his playground. Since there was a house for kids with mental disabilities in the neighborhood, she could teach her son how to deal with such people respectfully. As a result, she cherishes that he "has the open-mindedness to appreciate the diversity." She laments that the Highlands are not more ethnically diverse, but appreciates that gay, lesbian, and transgender people are accepted. She notes that for her suburban colleagues, parks are not enough because they are not private. For her, the fact that the park is full of people is a good thing; for the suburbanites, that would be a bad thing.

There are corporate class bourgeois bohemians, as well, who love the boburb.

A retired business owner was also enthusiastic about the collection of graceful older homes closely spaced on tree-lined streets, which he thought encouraged a sense of community. Sometimes people choose the suburbs over the city to be closer to "nature." On the contrary, though, he thought that typical suburban design is an environmental disaster, car-dependent, with artificially landscaped and chemically tended lawns and a nearly total absence of most wildlife. His Highland neighborhood, by contrast, had had time to recover from the damage of their development phases, and "the critters are coming back." In all, he concluded, "the older parts of the city tell a human story and the suburbs seem to be about developers making money."

The couple we met in chapter one, who said of their move from suburbs to Highlands "We knew our neighbors better within a year than we did in six years" in their old neighborhood, seem like natural suburbanites. As white, married, college graduates with two kids, who grew up in the suburbs, worked in business, and drove big pickup trucks, they themselves thought they wanted a suburban life. But after a few years they realized they did not fit in, precisely because they *didn't* want neighbors who all looked just like them. Moreover, they came to feel they had different values than many of their neighbors. They thought the norm in their old neighborhood was about "accumulating stuff" whereas they thought "more about experiences as opposed to stuff" – travel more, eat out more, shop less. They really like the variety of people who walk by their house now: "In the suburbs, when people are out at night, they [are assumed to be] up to no good."

A younger couple seemed even more primed for the far suburbs. They were religious, married, in Greek organizations in college, had private school backgrounds, played golf, served in the military, both had business careers, had a small child with more planned. And yet, they wanted a more urban experience. They "love the walkability" of the grocery store, farmers market, coffee shops, the park to run in, play with the kids in, go to summer concerts in. They are not cooks, so enjoy the abundance of restaurants. They are big advocates of public transportation. They have plotted out the sequence of public schools they hope their toddler will one day go to. They know they have different preferences than their friends, but think they are getting the better deal than suburban living.

An older professional couple, with many kids, who drove all the way out to a conservative suburban church on Sundays, decided they would either live in Lake Forest or the Highlands. They chose the Highlands because "our sidewalks go places." Sure, they say, if your neighbors fight, you know it; if your kids are bad, you hear about it. You hear dogs, you park on the street – "it's urban living." But they also love music, theater, arts, and go to everything.

"It's a huge sense of community here, I enjoy that. In Lake Forest, it wasn't like that – you parked in the garage; you never saw anyone. Here, you do know people." If you hollered, you would probably get some help. They appreciate the diversity of financial levels, educational levels, colors, people with kids, people without. "It's Halloween every day" – there is a tutu girl and a blue-hair boy. They like those old hippies who didn't want to follow the rules. They know that there are very few other Republicans in their neighborhood, but "it taught our children to love everybody. [They learned that] you don't look any different than anybody else. They're smart, too." It taught their kids to be in the world.

Our last couples are particularly interesting. They are married parents, active in their parish, active in Republican politics, both husband and wife with corporate professional careers – and they love the Highlands for the same reasons the artists and professors do.

In one case, he moved to the Highlands after college with fraternity brothers. They all enjoyed their bar-going years. As they married and had kids, though, all the rest moved to the suburbs as a matter of course. They think their frat brother is odd, if not irresponsible, for keeping his kids in the city. He does not find the city scary, and lets his young daughter ride her bike all over. What he does find scary is the Appalachian town he grew up in, which had a racism he did not want his children exposed to. Like other Highlanders, he likes walking to restaurants, talking to his neighbors, not having to drive to everything. He likes the fact that people "live on top of one another," which is a phrase suburbanites often use negatively. The social density of the Highlands means that "this area feels more like you are living in a larger city than most areas in Louisville."

The other couple had met in a large eastern city, but came back to Louisville to raise kids. She likes the fact that the Highlands is convenient and quirky, that they can walk to get coffee, that their house is old and has character. The oldness is important, they thought, for a strong sense of place; the old neighborhood, the big trees, the established families – "the families are as old as the houses." They know their neighbors, and praised the nosy dog walker who knew everyone's business. They had an interesting take on the value of diversity in the boburb:

> The mixed neighborhood fosters a more interesting culture. You know automatically that not everybody is the same as you. There's a balance of good families and people who are kind of like us and people who aren't like us, so it's not going to be too hard to instill our values. To get really snooty, it's more sophisticated than the east end in ways that I want to be sophisticated.

By aiming to "Keep Louisville Weird," the boburb, at its best, creates a form of sophisticated cosmopolitanism by celebrating the eccentrically local.

References

Jonathan Haidt, 2012. *The Righteous Mind: Why Good People Are Divided by Politics and Religion*. New York: Pantheon Press.

Misun Hur and Ashley Bollinger, 2015. "Neighborhood Associations and Their Strategic Actions to Enhance Residents' Neighborhood Satisfaction." *Nonprofit and Voluntary Sector Quarterly*, 44, 6: 1152–1172.

Allan Johnson, 2005. *Power, Privilege, Difference*. Second edition. New York: McGraw-Hill.

Part II

Bohemia and suburbia in Louisville

5 The Highlands and "the Glens"

The suburban alternatives

The automobile suburbs spread in great variety around Louisville. They range from the modest GI subdivisions of the 1950s to the McMansion gated communities of the 21st century. The car suburbs begin at the ends of the streetcar suburbs in the Upper Highlands or St. Matthews, then stretch to the very edges of Jefferson County, spilling over into the adjacent countryside. They were made possible by beefing up the rural pikes into paved, multilane arterial boulevards leading into the city, and by the expanding interstate highway system. While many people live in and love the suburbs, they typically do not have the same love for a particular suburb that boburbanites do for the Highlands. Thus in this chapter we will consider several neighborhoods as standing in for all many comparable car-based subdivisions. The Highlands are unique; "The Glens" are many.

The automobile suburbs

Today's suburbs were born when owning an automobile came in reach of the middle class. The government then shifted investment from rail lines – trolleys, interurban light rail, and long-distance passenger lines – to roads. The car gave more freedom to each family than a rail-bound vehicle could. The subsidized streets, roads, and highways gave the car a financial advantage that the private rail lines could never match. The shift began after World War I. By World War II most of Louisville's passenger rail options were dead.

The streetcars ran to the edge of the city. After 1900 new interurban lines ran small electric trains (what now would be called light rail) to the small cities out in Jefferson County, and to the seats of some of the neighboring counties. The interurban trains northeast along the Ohio River, and due east from Crescent Hill, turned some rich families' summer homes into year-round, commutable residences. Many of these little towns would incorporate as home-rule cities starting in the 1940s to fight annexation by Louisville. Some of these micro-cities are the richest places in Kentucky, especially Glenview, Anchorage, and Mockingbird Valley – a "city" of fewer than 200 people that has among the highest per-capita income of any municipality in the United States. The interurban lines would barely survive World War II, after which they were supplanted by automobiles.

The true car suburbs were built as master-planned subdivisions off the arterial roads, in the farmland between the interurban rail towns. Reachable only by car, they depended on massive government assistance to build the roads, and, through the GI Bill, finance the mortgages. The subdivisions excluded black families through racial redlining and restrictive covenants, even if they were soldiers who were (theoretically) eligible for GI bill mortgages. The returning white GIs and their young families flocked to the suburbs to raise the Baby Boom generation in miles and miles of single-family subdivisions. These are the classic "sitcom suburbs" (Hayden, 2003, 5).

Typical of these car suburbs is Barbourmeade.

In 1948 the first segments of the new perimeter road – what would become Interstate 264, the Watterson Expressway – were built in a semi-circle about eight miles from downtown Louisville. The country pikes intersecting this perimeter road were built up into suburban arteries. Subdivisions sprang up alongside these several arterial roads, near the highway interchanges.

Brownsboro Road was initially one such country pike. It runs close to Frankfort Avenue in Crescent Hill, before turning more to the northeast as it heads parallel to the Ohio River away from Louisville. It was, for many years, the main road from Louisville to Cincinnati, before it was superseded in the 1960s by Interstate 71. Barbourmeade was built on either side of Brownsboro Road, just outside the Watterson, in the 1950s. Part of its appeal was easy commuter access to both I71 and I264, highways which can carry a driver into the city or around it quickly.

The "city" of Barbourmeade was incorporated in 1962, though the only aspect of urbanity it contains are some 500 single-family houses, now holding about 1200 people. It surrounds Norton Elementary School, which is part of the Jefferson County Public School system, and is across the road from the Standard Country Club. The smaller houses are three-bedroom ranch houses of some 1500 square feet, selling for about $200,000, while the larger ones are two stories, twice the square footage, for over $300,000. The trees are well grown in. There are no sidewalks, though traffic is light enough that residents walk in the street safely.

The norm in Barbourmeade, as in most other subdivisions and home-rule cities, is that neighbors are civil with one another, but have a fundamental respect for each other's privacy (Baumgartner, 1988). The minority who wanted more intense, boburb-like interaction with their neighbors were often frustrated. The friendly couple, mentioned earlier, who set up their lawn chairs in their open garage in the hopes of talking to any passersby, lived there. A few others set up Next Door, a computer application that allows only those with a neighborhood address to message each other. While the organizers hoped it might foster broader sociability, it was mostly used, there and in other neighborhoods, to note local problems – fallen trees, potholes, missing pets, boats parked in driveways, stranger sightings – as well as for a handful of malcontents to fuss about things. There is no swim team, though some kids join the one in the adjacent subdivision; swim teams are often an important social connector in suburbs. Barbourmeade is a nice place to live – a "desirable neighborhood" in the real estate ads – but it is not very socially dense.

The families who choose other car suburbs tell similar stories. A young couple with a small child, both government workers, had wanted to live in the Highlands because "it is so much fun," but could only afford "dumpy" houses there. They live in a post-war ranch house, which was affordable and near their suburban workplace. Another couple with small children moved to the east end to be near her parents, though that meant a cross-town commute for him. A third couple had lived on a farm far from Louisville, but when they were expecting their first child, the thought of making a long commute with a little one seemed too dangerous. They bought a post-war ranch house with mature trees, chosen, in part, because it was near a top-rated elementary school. They have civil relations with their neighbors, and respect the woman who is the whole government of the tiny "city" they live in. All three of these couples are moderately liberal, but not very political. Showing political positions was not the norm in any of their subdivisions – there were few ideological stickers on cars, and no yard signs at all.

A couple of school teachers well illustrate the tension between their schooled understanding of what would be good, and their lived gut feeling about where they are comfortable. Both taught in the dense parts of the city, and wished they could engage their students in these neighborhoods more. Yet both had grown up in more rural places. When it came time to have children, they wanted a quiet cul-de-sac with yards that connected. They had taught near Bardstown Road, but did not want their own children hanging out there.

This couple also raised a concern often voiced in the suburbs: resale value. While most Americans move more than once in their lives, and all homeowners give some thought to the future value of their houses, this is an overriding concern in the suburbs. As public school teachers, they had among the most secure careers of anyone in the US. They did not have to worry about being transferred out of the city, and worried much less about being downsized out of a job. Yet, like many other workers, they bought with the expectation of leaving. And that expectation was not in the far future, when the kids were gone and they wanted a smaller house. Even the teachers, who could plan to stay and really liked their neighborhood, were ready to bolt for a better job or a better house. Thus, their choice of neighborhood was strongly shaped by their faith that the whole neighborhood, not just their house, would retain its resale value.

Resale value is of highest concern to the corporate managers who get moved by their employers every few years. From the 1950s onwards, the Louisville economy has been dominated by large national corporations. These firms either bought up the local businesses or built new plants of their own (Yater, 1979, 222). The management class has the most money to spend on housing, and the most interest in neighborhoods with easy-to-maintain new houses that will resell quickly. The new car suburbs, especially the fanciest ones, were made to order for this new economic reality.

Which brings us to one of the overriding distinctive features of suburban subdivisions, the powerful legal structures to maintain property values.

The great majority of single-family houses built in the last two generations have been made in "master-planned subdivisions." Rather than a family buying a plot of land and building their own house, as was the norm a century ago, developers buy whole farms, subdivide it into quarter-acre lots, plan the streets and the utilities, and build a few standard model houses. The term "master-planned community" is sometimes used by real estate dealers and zoning nerds, but is not in everyday use. "Subdivision," on the other hand, is a common way that suburbanites talk about their neighborhoods.

A legal innovation that came with these post-World War II subdivisions is the use of Covenants, Codes, and Restrictions (CCR) included in the deed, which are explicitly designed to protect property values. Early covenants had egregiously racist restrictions on ever selling to Catholics, Jews, or, especially, African Americans. These restrictions were invalidated by the courts. However, deed restrictions on the appearance of the property have been allowed. These include the colors the house can be painted, the kinds of landscaping allowed, the types of vehicles that can be parked in the driveway, limits on renovations that can be seen from outside, how long the grass can grow, and severe limits on art, banners, posters, or signs that can be displayed.

When a subdivision is being developed, enforcing these rules is the builder's responsibility. As soon as the neighborhood is populated enough, though, the developer passes on – sometimes thrusts away – the enforcement duty to a Homeowners Association (HOA). The HOA is a voluntary group, normally elected by the residents. It is, however, so hard to get people to serve on these boards and enforce the common rules on neighbors that often it is a dutiful few who get coopted on to the board, then later elected. The HOA charges dues, can levy fines, and, in extreme cases, can put a lien on a property – even foreclose for failure to follow the rules. HOAs sometimes employ a management company to maintain the common areas, do the legal paperwork, and advise them on larger projects. The HOA is, in effect, a form of local government. One longtime HOA president had a more cynical reading of why this form was created: "HOAs are nothing more than a way for local government to avoid enforcement of things that need to be enforced." He further offered that CCRs were invented to protect the developer from long-term liability.

The CCR and the HOA are among the most distinctive features of today's American suburbs. They were an alternative to creating the tiny home-rule cities, like Barbourmeade. Counties favored either structure, because it passed on the responsibility for providing some services, especially trash collection, street lighting, snow removal, common-area plantings, and, in some cases, security. By contrast, residents of the boburbs have much more freedom about what they can do with their own property, and with the neighborhood as a whole. The deed restrictions are why you will almost never see a purple house in a suburban subdivision.

Suburban subdivisions usually have recreation facilities for residents, from modest playgrounds, through volleyball courts, tennis courts, swimming pools, even whole golf courses. In many suburbs today the neighborhood swim team is the most shared social institution, the main tool for building community solidarity and personal connections among parents and children. It requires continuous labor by the HOA and the club members to physically maintain a swimming pool. Just as important, it requires continuous social labor to recruit and maintain a team of young swimmers, each of whom rapidly ages out of eligibility. If the motto of the boburb is "Keep Louisville Weird," the motto of the suburb might be "Keep the Swim Team Full."[1]

The gated communities

Already in the 1960s the Inner Loop – the perimeter road that would become the Watterson expressway – was so heavily congested, especially in the booming eastern suburbs, that planners started on an Outer Loop. Construction even began on a short connector between Interstate 64, the main road to Lexington, and other eastbound pikes and arterial roads. Local opposition, especially from richer communities, slowed construction for decades, though. It was not until the late 1980s that the main Kentucky portion of the highway was completed. This interstate-grade highway connected I71 in the northeast to I65, heading south to Nashville. Congressman Gene Snyder was so influential in getting federal funds for what became Interstate 265 that the road was renamed for him. It is usually called "the Snyder" by locals. An Indiana portion began at I64, west of Louisville, and connected to I65 north of the city, heading to Indianapolis. Finally, in the 2010s, road builders were able to work around the richest communities in the state, whose estates were northeast of Louisville along the Ohio, to build a new bridge and the approaching highway to complete the northeastern arc of the circle. This outermost perimeter road is now three-quarters of a circle, from 9 o'clock to 6 o'clock on an analog clockface.

The Snyder fostered subdivisions even further out into Jefferson County. Among the most ambitious is Lake Forest. In 1981 plans were announced for a massive, 450-acre luxury subdivision just east of this new perimeter highway and north of I64. Sales were brisk enough that Lake Forest claimed to be "Louisville's most successful residential community," located some 15 miles from downtown. A 1985 newspaper ad, complete with dramatic ellipses, contended:

> You've been looking for the good life … a lovely home set in privacy in eastern Jefferson County. Surrounded by cool inviting lakes. Wooded lots. And a gracious lodge for entertaining. You've worked for it. And now you've found it … at Lake Forest.

To drive home who their target market was, the open-house offer concluded "Bring your family, and we'll provide the babysitter" (NTS Builders, 1985).

In the fashion of such luxury developments, Lake Forest was a "gated community," a phrase that first appears in the *Courier-Journal* in real estate ads for The Springs subdivision many times in 1996. This meant it had impressive stone gateposts at the entrance to Lake Forest Parkway, the main road. There is not, though, an actual gate, nor are there guards, though such communities often have official-looking guardhouses. Gated communities were first developed for retirees in the desert Southwest, who wanted to feel completely safe. The idea came late to Louisville because, the builder of Lake Forest reasoned, the crime rate was so low in the Jefferson County suburbs. However, he thought the market for gated communities would grow "as more people who are used to living behind gates in other areas move into this community" (Bennett, 1998).

Lake Forest is the subdivision that real estate agents first mention – off the record – if you ask for "McMansions," a word that also first appears in the *Courier-Journal* in 1996 (Walfoort, 1996). McMansions are houses of 3,000 square feet and much more, set close together on relatively small lots. In addition to the lodge, swimming, tennis, and other recreation facilities, in the 1990s Lake Forest added a new section, along Arnold Palmer Drive, around a golf course designed by the famous golfer himself.

Managers and, to a lesser extent, professionals with more money to invest in housing, move to Lake Forest. The business plan for McMansion neighborhoods is to appeal to executives who get transferred often, who are looking for zero-maintenance houses with high, quick resale value (Hershberg, 1981). People move in late in their careers, when they can afford the expensive houses and HOA dues, then downsize a decade or so later when the children move out. Gated subdivisions are not really intended to be long-term communities. This neighborhood draws people with more conservative politics, and a greater desire for security, than just about any other place in Jefferson County. Reflecting the conservative social views of the target audience, Arnold Palmer said:

> We'll build a course that on Sundays you can bring your wife and children and have a very enjoyable afternoon. But on Wednesdays with a foursome of scratch golfers, you'll play a course that's enough of a challenge that you'll wonder why you enjoyed it so much on Sunday.

Gated communities are aimed at protective patriarchs of straight families with children (quoted in Terhune, 1989).

Lake Forest is in some respects the antithesis of the Highlands. It is meant to be a country club – emphasis on the *country* – community for just one class, literally walled off from the jangle of the city.

Visitors to the fanciest gated communities notice the big houses amidst the immaculate landscaping and the elaborate recreational facilities. Some golf course neighborhoods weave the course in between the houses, around the private lakes. Yet what the residents tend to talk about are not the material riches of their neighborhood, but safety, privacy, and a sense of control over their space. One long-time resident of a golf community encapsulated the virtues of his home locale as a "safe, clean, place you could invite your friends to, where somebody is watching over what is going on."

One couple well exemplified the kind of family most at home in the gated community. Married, white, with several children, they had come from small towns. His corporate career had moved them all over the country. Each time they chose neighborhoods like Lake Forest. They wanted the safety, the swim team, and the secure resale value. They describe themselves as very conservative, attended a conservative church, and sent their children to Christian schools. He said he did not feel especially attached to their current gated community, but he did serve on the HOA in order to be sure that the rules were enforced to protect property values.

Several conservative political professionals perfectly exemplify the gated community demographic. When they began their housing search, each said they only looked in the moneyed suburbs – "east of the Watterson, north of I64." They wanted bigger houses, more land between them and the neighbors, private amenities such as golf courses or private lakes. Most of all, they wanted to control who they interacted with. The one area in which this control might be uncertain was in schooling, which they solved by either assuming they would choose a private school, or moving over the county line, out of the Jefferson County school assignment system. These married white fathers were explicitly choosing safe, clean places for their families, without articulating what, exactly, they were keeping their families safe from.

In fact, it was striking that, while all the gated community residents appreciated where they lived as a *space*, almost none were attached to it as a distinctive and beloved *place*. They all planned to move away, especially when the kids were grown. They expressed a desire for even more privacy, less responsibility, more leisure – beach, lake, golf course. None expressed a desire to have their children come live near them in their current neighborhood. None expressed a desire to live somewhere with more community, more solidarity, or more social connection.

I talked to two experienced professional managers of deed-restricted subdivisions, both as on-site managers in richer, larger neighborhoods, and as outside managers working with the HOA for a portfolio of smaller communities. They estimated that 80% of new construction in Jefferson County is in deed-restricted communities, including infill in older areas. Indeed, they guessed that half the housing in the east end of Jefferson County was already under CCRs, and unregulated areas are "becoming extinct." In their experience, most people who live in an HOA neighborhood like it. Residents believe the association will act on the neighbors who do prohibited acts, such as having a boat in the driveway, because the residents "feel that their [property] values go down if you let those things happen." The managers asserted that deed-restricted neighborhoods sell for 20–30% more than other properties.

The managers offered a nuanced account of the kinds of social exclusions that Louisville's deed-restricted communities create today. On the one hand, they found status distinction to be pervasive. The east end was "the place to live," and some people buy more than they can afford because they want that status – "They consider themselves a little higher than other people." These women were conscious that country clubs, which exist symbiotically with master-planned communities, "don't take women seriously." On the other hand, they noted that when a gay couple in a tony neighborhood was featured in a non-community magazine, the manager received complaints, but the subdivision's professional managers supported the couple. Perhaps most importantly as a sign of changing mores, they could not really grasp the concept of racial exclusion of families that could afford to live in any of the communities they managed.

The managers themselves were divided on whether they wanted to live in a CCR community. One did, because she wanted her property values kept up; also, her husband wanted to live near a golf course. The other manager, though, chose not to live in a deed-restricted community because she didn't want the restrictions. Both women said they enjoy people watching on Bardstown Road. In fact, they allowed, if they could live anywhere, they would both prefer to live in the Highlands.

Note

1   I am grateful to Stephanie Keller for this formulation.

References

M.P. Baumgartner, 1988. *The Moral Order of a Suburb*. New York: Oxford University Press.

Doug Bennett, Jr., 1998. "Gated Communities Offer Security, Privacy." *Courier-Journal*, April 4, Home Showcase 4.

Dolores Hayden, 2003. *Building Suburbia: Green Fields and Urban Growth, 1820–2000*. New York: Vintage.

Ben Z. Hershberg, 1981. "450-acre Luxury Subdivision Planned in Jefferson by NTS." *Courier-Journal*, October 31, Marketplace B8.

NTS builders ad, 1985. *Courier-Journal*, July 26, C11.

Jim Terhune, 1989. "Palmer vs. Nicklaus. Arnie Designs Course Next to Valhalla." *Courier-Journal*, May 25, E1.

Nina Walfoort, 1996. "Critics Assail Mundane Suburbs." *Courier-Journal*, April 4, B3.

George Yater, 1979. *Two Hundred Years at the Falls of the Ohio: A History of Louisville and Jefferson County*. Louisville: The Heritage Corporation of Louisville and Jefferson County.

6 Germantown and Clifton

Bohemia the old-fashioned way

The Germans settled the eastern edges of Victorian Louisville, establishing truck farms, breweries, and slaughterhouses. The swampy ground along Beargrass Creek had not been claimed by earlier settlers, so they created a string of neighborhoods along the creek. In the 1870s when the creek was better controlled, textile mills were built over the old farms, and rows and rows of shotgun houses were built for their workers, still mostly German. Today several of these neighborhoods between pre-1830 Louisville on the north and west, and the Highlands on the east, are collectively known as Germantown. The core neighborhoods are Germantown proper and Schnitzelberg. Adjacent Meriwether is often considered part of Germantown now, as is the small triangle called Paristown Point, the residue of a French Catholic neighborhood.

On the far, eastern side of Beargrass Creek, German butchers built Butchertown. It now merges into Clifton. In the 1840s one of the rich men of the city, Col. Joshua Bowles, president of the Bank of Louisville, moved to the bluffs near Frankfort Pike, and named his estate Clifton. Gradually, an industrial area grew around the base of the bluffs extending to the river. On the eastern side of Clifton the land makes a quick rise of some 20 feet above the flood plane, then begins a gradual climb to Crescent Hill, two miles further east. This higher ground and country air drew the campuses of several philanthropic institutions. The first was the Kentucky School for the Blind, which moved from downtown to Clifton in 1855, and is still the dominant institution of the neighborhood. Louisville annexed most of Clifton in 1856 (Rogers, 1955, 8).

"Germantown" as a *concept* is coming to mean the former working-class neighborhoods into which young college graduates are moving. Butchertown and Clifton, though not technically part of Germantown, are enjoying a similar shift from their previous working-class identity. Shelby Park, which is predominantly black, is the front line of the change – gentrification, bohemianization, renewal – that is moving through all these formerly working-class neighborhoods.

The Germans who settled Louisville from the 1840s to after the Civil War were Catholics, Protestants, and the first communities of Jews in the city. The

German Protestants melted into the broader English-speaking Protestant community in the city, and the Jews moved as a group to Old Louisville and the Highlands before dispersing into the car suburbs. Many German Catholics, though, stayed in their several Germantown neighborhoods, near their parish churches and factory jobs. This is why Germantown remained white when many of the adjacent working-class neighborhoods, such as Shelby Park, became predominantly black in the 1960s and 1970s.

The distinctive housing style of Germantown, the shotgun, is another reason why it was not gentrified before – but is ripe for bohemianization now. Shotguns are small – typically 12 feet wide, perhaps 80 feet deep, with a series of rooms opening one into another front to back. The fancier ones have a side hall, and perhaps a two-story "camelback" addition in the back. In Butchertown many were made of brick, but the great majority in Germantown are wooden. Shotguns were a popular working-class housing style from the Civil War to World War I; most of Louisville's were built in the 1890s, when the streetcar line was extended to Schnitzelberg. Louisville has so many, in fact, that shotgun neighborhoods have become a cultural feature and tourist landmark.

Shotguns are also small enough that they remain relatively cheap. Whereas the houses in the Highlands were built for the middle and upper-middle classes, Germantown was built for the working class. The 1,000-square foot shotgun on a narrow lot will always be cheaper than even the 1,500-square foot bungalow with a second floor and perhaps a garage.

Shotgun neighborhoods have several advantages for sociability. The house is built close to the street. If the house has a porch, or even a stoop, hanging out on the front porch of an evening, conversing with the passersby – or even the next-door neighbors – can be done without shouting. German workers developed the biergarten as a walkable social location for the whole family, and the working-class taverns dotting Germantown keep up the custom into the 21st century.

Germantown – and most of the neighborhoods within the limits of the streetcar city – have an advantage for free spirits over nearly all the car suburbs: no aesthetic restrictions on your property. Homebuyers in today's deed-restricted car suburbs have agreed to limit the colors of their houses, the objects in their yards, the vehicles they can park outside, and many forms of expression of opinion. Germantown was built long before such restrictions were invented. Bohemians love shotguns, and shotgun neighborhoods, because they can put art in the yard, put up any opinion signs they want, and treat their wooden houses as a large canvas for self-expression. The renovated shotguns are marked by a wide array of colors in wild combinations. There are frequently sculptures in the yard, made by the inhabitants or their friends. In bohemia there is always a purple house.

The decline of Germantown as a working-class neighborhood, and its renewal as a bohemia

Germantown was a solidly working-class neighborhood as long as there was work. The Louisville and Nashville Railroad (the L&N), a dominant firm in Louisville in the 19th century, ran through Germantown. Its tracks and rail services drew several factories there. In the 1970s, though, the railroads were being displaced by cars and trucks, which ran on tax-subsidized streets and highways. The L&N was bought by larger railroads, and was fully absorbed by CSX in the 1980s. The factories closed and stood vacant. The workers moved away if they could, or lived on pensions in their paid-for houses near their old parish churches.

At the same time, the Highlands was beginning its transformation from affordable, down-at-the heels bohemia to too-expensive, more family-oriented boburbia. The young artists, bohemians, and hipsters of the 21st century started to pass up the Highlands as too expensive, too bourgeois, or both.

This then pushed the would-be bohemians to develop other areas. Since the turn of the 21st century, the shotgun houses of Germantown have become the "new Highlands," where young people with more education than income first move to live an arts-and-culture life. Germantown is the new bohemia of Louisville. The dive bars are being displaced by craft-beer taverns. A coffeehouse has opened, serving espresso and Vietnamese coffee. Liberal bumper stickers have multiplied. Pickup trucks with college stickers line the street. And amid the tidy small houses maintained by the remnants of the old working class appear houses painted bright colors with odd sculptures in the yard.

Perhaps the best mark of a bohemia is a proliferation of art-making spaces. A journalist who covers both arts and the craft of distilling called German-town "the new Highlands." She cited Art Sanctuary, a "community oriented arts collective" in an old warehouse on the edge of Schnitzelberg as a site for many kinds of art-making (Art Sanctuary, 2018). The open studio tour orga-nized by the Louisville Visual Arts Association had 22 studio stops in Ger-mantown (Louisville Visual Arts Association, 2018). Louisville's major popular music event, the Forecastle Festival, is a huge event for national acts, held downtown on the riverfront. Its counterpart for local musicians is Poor-castle, a "Festival for the Rest of Us," held at the Apocalypse bar in Clifton (Poorcastle Festival, 2018).

A good example of the rise, fall, and re-purposing of the industrial struc-tures which made Germantown work was the Louisville Cotton Mill. The company built a large factory in 1889, and did a steady business through World War II, employing hundreds in the immediate neighborhood. By 1960, though, the mill had closed. The space was partly abandoned, and limped along as a flea market. In the 2010s it was refurbished as Germantown Mills Lofts, offering $1,000 per month apartments. The bohemians, who had been buying the little shotgun houses from retired workers, view these lofts as a yuppie intrusion that would displace the bohemian community even as it was being born.

Who is moving to the new bohemia? All the people I interviewed liked the things that their counterparts liked about the boburbs. They appreciated the walkability, especially of the bars. They liked that these neighborhoods, which are even closer to downtown than the boburbs, were "near everything." They appreciated the diversity compared to the suburbs – "you don't see yourself coming and going" in every other face. Moreover, because the bohemianizing neighborhoods are built on a working-class base, there is significantly more class diversity than in any other neighborhoods in town. The newcomers tend to be very liberal – "royal blue" in the words of one neighborhood association leader – even as the old residents tend to be very conservative. Gentrification was a concern for some of the liberal newcomers, but this was tempered by the fact that in Germantown, at least, the displaced were dying off more than they were being driven out.

Boburbs don't grow automatically, but are the fruit of active local organizations, governmental and otherwise, building up the community. Any poor or working-class neighborhood might be attractive to bohemian artists. However, Clifton is more likely to become a boburb than other working-class neighborhoods with cheap housing because the Clifton Community Council has been effective. Activism by newcomers and some long-time residents led them to be named a "preservation district" in 2003, which has helped the council and city officials to stabilize the neighborhood. The council created parks, encouraged local festivals, promoted "buy local" campaigns within the neighborhood, and paid special attention to incorporating the many blind residents. The council promotes political engagement, which in turn fostered a general sense of efficacy for Clifton as a place (Stephenson, 2017).

Clifton is a more bourgeois bohemia than is Germantown, just as adjacent Crescent Hill is a more bourgeois boburb than the Highlands. Middletown United Methodist Church opened a street ministry, You First, on Frankfort Avenue, in the mid-2000s. Middletown is a very suburban town, outside the Watterson. The Asbury seminarian they hired to run the program allowed that the church might have some misconceptions about the people it would reach when they decided to open the mission in Clifton. The church "initially expected to attract young, tattooed and pierced bohemian types, but what it has found is very different. The majority of people living in the area are in their 20s and 30s." An increasing number have young families (Hall, 2006).

Some natural boburbans lived in Germantown because, despite rapidly rising prices, bohemian shotguns were still cheaper. A social work student, working for a non-profit, asserted that "everyone would like to live in the Highlands or Crescent Hill," but she could not afford to live there on one income. A married couple, both graduate students, wanted a yard for the dog. They like their shotgun, but are probably not there for the long haul. A married couple of liberal activists really enjoyed their neighborhood when younger. Now that they are a bit older and have a small child, they don't feel "hip enough" for Germantown, and anyway she worries a bit about "randos" walking by their house at night. They have made an offer for a house east of Crescent Hill; since they are still inside the Watterson (barely), though, don't feel they have "sold out" and become fully suburban.

Still, the culture of bohemia was a draw for some people who thought the Highlands and Crescent Hill were not bohemian *enough*. Some are classic bohemians, the artists and musicians who created the stereotype of such neighborhoods in the first place. One man grew up in the country, went to college, and became a roadie for rock bands. When he met a woman who made costumes, he settled down as a sound engineer for Louisville musicians and theaters. They live together in a house in Germantown. They have no plans for kids. They like walking to their favorite bars, and being near the arts venues. They think the Highlands has become "too yuppie."

A new kind of bohemian is the college-educated "masters of craft" (Ocejo, 2017). A woman who graduated from a top-ranked liberal arts college, then worked in Europe, developed a taste for craft beer. She moved to Louisville to get in on the ground floor of a new microbrew pub. She now works at three bars in the bohemian quarter, and loves Germantown. She said that the Highlands was just too expensive, and the residents too old for her. At the same time, the established bars on Bardstown Road drew suburban kids and college students who "came to get really drunk on weekends." She said most of her Germantown friends are young college graduates, "hanging out and figuring it out," who are enjoying their late-night youthful explorations. She pointed out that most carry quite a bit of student debt, which is easier to pay off when living in Louisville than in a more expensive city, like New York or Chicago. She does not like to go to the malls or chain stores, and never goes out to their suburban neighborhoods if she can help it.

Some people who were not straight, white, married parents – and never would be – just did not feel fully at home in the suburbs. They reported very few incidents of outright hostility or exclusion by other educated people, but they felt themselves to be not part of the hegemonic norm. They wanted a neighborhood that was more diverse, in part because *they* lived there. A white couple, who grew up in the suburbs, met in college, got corporate jobs, who might otherwise have spent a few years in Crescent Hill before kids sent them beyond the Watterson, chose Germantown because, as lesbians, they just felt a little more comfortable. A professor living with a partner likes the walkable bars and the affordability and favored gentrification – "renovating dumps is a good thing." What sold her on buying in Germantown was that they planned never to have children, so they could reap all the benefits of bohemia without one of the chief worries of people like them. A married lawyer was raising his kids in the Highlands; when he divorced and did not have primary custody of his children, he felt it was safe to move himself to a more dangerous bohemian neighborhood and throw himself into its renewal.

The more factors that removed one from the white middle-class family norm of the suburbs, the more bohemia seemed the right place to live.

Likewise, people with "diversity promotion" careers found the suburbs too "white bread" to enjoy. A white woman deep in a social service career lived with a fiancé in the far eastern suburbs, until she concluded that she could not find real community there. She broke up with him and moved to Clifton. A white lesbian couple with children, both in social service jobs, worked with suburbanites, but didn't want to live there. Two single non-white women each moved in from the suburbs to have a more fun life as confirmed single homeowners. An interracial married couple, with no plans for children, did not want to go back to the suburbs they grew up in. A black couple who study racial disparities, new to the city, tried semi-suburban St. Matthews first. They never encountered any ugliness there, but when they went into a bar with a "no gang attire" rule and blaring sports televisions, they knew they were not quite the target demographic of the area. An interracial trans couple were first steered to the suburbs because of their professional credentials and income. They did not face any direct unpleasantness, but were treated as notably unusual. They now laugh that they thought they would feel at home in the suburbs.

A black professional, who grew up in the poor black west end and has a career in urban renewal, did not want to live in the west end. At the same time, she would have felt like a sell-out moving to the east end. She chose a traditionally black neighborhood that is now bohemianizing as the right middle ground. She is well informed about the history of redlining that restricted where African Americans could live. She also appreciates that bohemianization is both gentrifying, in the bad sense, but is also undoing the effects of redlining, in a good way.

References

Art Sanctuary website, 2018. http://www.art-sanctuary.org/.

Christopher Hall, 2006. "Center Reaches Out to Clifton Residents." *Courier-Journal*, March 4, B2.

Louisville Visual Arts Association website, 2018. https://issuu.com/louisvillevisualart/docs/2017_osw_directory_final-web5.

Richard Ocejo, 2017. *Masters of Craft: Old Jobs in the New Urban Economy*. Princeton: Princeton University Press.

Poorcastle Festival website, 2018. https://www.poorcastle.com/.

John Rogers, 1955. *The Story of Louisville's Neighborhoods*. Collected stories from the *Louisville Times* of May 1955. Published by the *Courier-Journal* and the *Louisville Times*.

Aaron Stephenson, 2017. "Norton Commons, Clifton, and Social Equity: A Neighborhood, Inter-Neighborhood, and Regional Comparison of New Urbanism and 'Old Urbanism.'" University of Louisville PhD dissertation.

7 NuLu and Portland

Bohemia on steroids

In the 21st century there is an ambitious experiment to lift two faded industrial districts into hip arts and entrepreneurship centers. These instant bohemias may accelerate the process that shaped the Highlands from a more traditional starving artist bohemia in the 1970s to the thriving boburb that it is today. There is no guarantee that this revival will work, nor that it would turn into a boburb down the road. But the forces behind this project have taken a lesson from the movement to lure the creative class to the city through vibrant cultural scenes.

NuLu

Market Street was, as the name suggests, the main commercial street of Louisville from its earliest days. The eastern end was marked by the wet lowlands around Beargrass Creek, behind which lay the first rise that would lead to Crescent Hill. The Bourbon Stockyards grew at that spot, with the offal of the slaughterhouses dumped into the creek. This smelly neighborhood became known, inevitably, as Butchertown (Kubala, 2010). The ten blocks between the stockyards and 1st Street became the core shopping district of the city in the 19th century, and well into the 20th.

In the 20th century East Market, as it came to be called, had already started declining as a retail destination, displaced first by 4th Street, and then, after World War II, by the suburbs. This area was still important for wholesale and light manufacturing, though, until the 1960s and 1970s. The general economic decline of the old parts of the city was accelerated by the mass departures of the middle class, and their businesses, after the race riots and busing controversies. The stockyards closed in 1999, ending an era. East Market became known mostly for social service agencies for the worst off, such as Jefferson Street Baptist Center, Wayside Mission, and the Home of the Innocents children's home, which took over the old stockyards space.

Unlike all the other neighborhoods we meet in this study, though, the East Market district was never much of a residential neighborhood. Some workers resided in German Butchertown to the east, and many more in Irish Phoenix Hill or black Smoketown to the south. The owners lived in Old Louisville at first, then the Highlands or Crescent Hill. When the businesses declined, many commercial buildings just stood empty for years.

In the early 21st century, though, a concerted effort is being made to revive the eastern end of Market Street. The moving force is arts entrepreneur Gill Holland and his wife, Augusta Brown Holland, an heir of the Brown-Forman liquor fortune. Augusta Holland supplied the intellectual grounding for their joint project. She was the one who had studied Jane Jacobs' urban theories and Richard Florida's creative class theories, and who had a commitment to sustainability (Fleischaker, 2016). The Hollands were informed by the conviction that historic preservation of old buildings is smart, that environmental sustainability is necessary, and that the arts could be a catalyst for other economic development.

In 2007 the Hollands bought depressed commercial property in East Market. They renamed the area NuLu, and turned a vacant commercial building into their own Green Building to show all their convictions at once: that an historic building could be restored in an environmentally sound way to house an arts business that could draw other creative industries. The Green Building claims to be Louisville's first commercial Platinum LEED certified building, and Kentucky's first Platinum LEED adaptive reuse structure. The building is home to an art gallery, the headquarters for sonaBLAST! Records and Holland Brown Books, and La Coop café (Kolleeny, 2010).

Gill Holland began his career as a filmmaker, and had lived in the iconic American bohemia, Manhattan's East Village, before moving to his wife's hometown of Louisville. In 2012 Gill Holland told the origin story of the consortium to remake East Market this way:

> One day about three and a half years ago, signs were posted saying that the Wayside block of 150-year-old buildings was to be imminently demolished. Many folks suddenly realized that, in the Jane Jacobs spirit of "new ideas need old buildings," this would not be in the best interest of Louisvillians. We hurriedly put together an investment group and Wayside agreed to be purchased for a nice profit to them. They were also given two years free rent, until they were able to close the deal on the Holiday Inn on Broadway. In retrospect this private deal has been a great win-win for the community. No public money or subsidies were used at all to transform a federally designated economically distressed area into a booming arts, cultural and sustainability district, arguably the coolest such area in the Mid-west.

The appeal to Jane Jacobs, the patron saint of the new urbanism, is there at the start. To the old bohemian ideal of an arts and culture district he adds the new concept of a "sustainability district." The idea that the arts, in renovated "cool space" in old buildings, can drive a city's economic development, is an idea spread by Richard Florida to many city leaders throughout the world – including in Louisville (Florida, 2002).

The Hollands and their partners drew other entrepreneurs to open their own stores, including galleries, coffeehouses, pubs, clothing stores, breweries, and an abundance of restaurants. Convinced that what Millennials need is "coffee and wifi," Holland offered the new Please and Thank You coffeehouse free rent to open on a corner in the heart of the district. They also convinced arts organizations, such as the Louisville Ballet and Sarabande Books, to move from their long-time home in the Highlands to newly upgraded space in NuLu. The business development was heavy on new mom-and-pop entrepreneurs, rather than stealing established businesses; chain stores were not even on the menu. The business leaders also saw that the richness of the neighborhood would be strengthened if they saved the local elementary school. Lincoln elementary was academically declining, had almost entirely poor students, and was headed to closure. The NuLu entrepreneurs led a drive to restore the school and redirect it to be a performing arts magnet elementary school, the only one in the city.

The Achilles heel of NuLu as a boburb, or even as a bohemia, is the lack of housing. There never was much housing in the immediate area in its earlier commercial days. One developer was renovating commercial buildings one at a time, with business at the bottom and a "very elegant" apartment above, but this provided very few new residences (Walfoort, 1999). With the revitalization of NuLu and the waterfront areas just east of downtown, some big luxury apartment buildings have been constructed. The city has pressed for new projects to include at least some "workforce or affordable" housing in new apartment construction within NuLu. For example, the developer of one 270-unit building agreed to set aside six units at a lower rate, and contribute to a fund to build affordable housing – somewhere else in the city (Shafer, 2017a).

The experience of an old resident and a new one is revealing about who wants to live in NuLu, and what it is like to live there. An artist who grew up in the dense part of the city started a business there in the 1970s, even as many middle-class people were leaving town. When he married and had a child, they first lived in the country. However, they decided they wanted to live in the city and desired urban opportunities for their kids. They knew how to live in a city, from living in Louisville and from their experiences of New York and Paris. They moved into what would later become NuLu at the depths of its decline. Both of them lived and worked in the core city, walked to work, happily enrolled their kids in progressive downtown schools. They walked or biked to concerts on the waterfront, plays at Actors Theater, baseball games at Slugger Field. They found a liberal church, in part to find other children, who were in short supply in NuLu. Their more suburban friends and relatives worried about crime, traffic, and dirt, which he answered with statistics showing that his neighborhood was safe as the suburbs and had less traffic – except when the suburbanites were coming and going.

Another couple reacted against the fearfulness of their relatives in the suburbs and embraced the diversity of the city. They first lived in Clifton, then moved to a suburb inside the Watterson when their child went to school.

Their child's move to college coincided with their parents' final passing, and the massive project of dealing with all of their parents' stuff. This made them decide to downsize dramatically. From their experience in France as well as in Louisville, they preferred a vibrant downtown to the country. They moved into a new high-rise condominium in NuLu. The condo building is very diverse in age, sexual orientation, race, and ethnicity – "an old hippy's dream in terms of diversity and camaraderie in the building." They wanted more community connection than they had had in the east end, so they joined the condo board to help make events happen. They go to neighborhood meetings, go out to evening events, and "do all the festivals."

NuLu has clearly taken off as a hip commercial neighborhood. If it can find an affordable place for its culture-makers to live nearby, it has potential as a future boburb.

Portland

If you head west on Market Street from NuLu, passing through the heart of downtown, you soon reach Portland. From 12th Street westward the area north of Market Street all the way to the Ohio River is the current incarnation of what used to be the separate city of Portland. Since the 1840s, Portland has been incorporated as a working-class neighborhood of Louisville. Since the 1970s it has been the poorest white neighborhood in the city. It sits across Market Street from Russell, one of the poorest black neighborhoods in the city.

Revitalizing Portland has been the Hollands' next project after NuLu. Their success in NuLu made it easier to recruit other entrepreneurs and non-profits to join in the project. In the 2010s their Portland Investment Initiative had put tens of millions of dollars into revitalizing Portland (Portland Investment Initiative, 2018). They bought old warehouses by the acre (Shafer, 2017b). They worked with the government to extend the riverside park. They convinced other Louisville businesses to locate there. One of the central symbols – and players – in boburban culture in Louisville, Heine Brothers Coffee Company, refurbished a large Portland warehouse as their roastery, training center, and corporate offices. Galleries, coffeehouses, and clothing stores followed. The University of Louisville moved its fine arts department there, and may build housing for art students (University of Louisville, 2018).

The Portland project continues the NuLu emphasis on historic reuse, arts-heavy economic development, and environmental sustainability. The goals of the investment group add a new phrase to the list: "bolstering the neighbor-hood's intellectual infrastructure." Though Holland and his partners are pri-vate, for-profit "impact developers," they deliberately take lower and slower profits in order to develop the community. He says he wants Louisville to be "the coolest city in America" so his children will want to return to their hometown after an assumed sojourn in the broader world. Walkability is an important part of mixed-use neighborhoods, and is vital to what makes a boburb flourish. The Portland Investment Initiative includes a whole section devoted to a "Portland Stroll District." Their bumper sticker: "Portland: Louisville's Last Bohemian Neighborhood."

A great advantage that Portland has over NuLu as a bohemia now, and possibly a boburbia in the future, is that it has always been a residential neighborhood. Even better, it has been composed of small, detached houses – ideal for young people and starving artists of all ages. In addition to promoting commercial development, the Hollands have been drawing in apartment developers, and putting money into renovating the small shotgun houses that were the backbone of working-class and petite bourgeois homeownership in Louisville for a century. The Portland Investment Initiative will not only renovate old shotguns, it will work with architects to develop "21st century" shotguns to be constructed in the many vacant lots.

Right now, college graduates who move to Portland tend to have special circumstances. One couple with several children made a conscious decision to return to their Portland roots after college. Both work in social services. Their church was deliberately located in Portland with a revitalizing mission. They intentionally live very frugally. Their only worry is the quality of education their children might get, but they also know well how to navigate the school district's assignment system. They are not afraid of living in a poor neighborhood with a higher crime rate. As a white family, they fit a crucial local demographic.

A man who lives on the border between white Portland and black Russell is also an unusual case. A single black man, he felt he had fewer security worries than he would if he had a family to think about. An artist, he appreciates the grit of the city. As an entrepreneur, he appreciates the low cost of property in what is, at the moment, a slum. He sees, though, that the development of Portland, and the possible development of adjacent neighborhoods, is likely to improve the value, and security, of his home.

References

Greg Fleischaker, 2016. "Perspectives: Redevelopment of NuLu with Gill Holland." Blogged August 18. https://www.lenihansothebysrealty.com/blog/perspectives/re-de velopment-of-nulu-with-gill-holland/.

Richard Florida, 2002. *The Rise of the Creative Class: And How It's Transforming Work, Leisure, Community and Everyday Life.* New York: Basic Books.

Gill Holland, 2012. "NuLu's Culture as an Economic Driver." *The Voice-Tri-bune,* April 18. https://voice-tribune.com/x-unused/your-voice-news/nulus-cul ture-as-an-economic-driver/.

Jane Kolleeny, 2010. "Case Study: The Green Building." *The Green Source,* September. https://web.archive.org/web/20101010023752/http://greensource.construction. com/green_building_projects/2010/1009_Green_Building.asp.

Edna Kubala, 2010. *Louisville's Butchertown,* in the Images of America series. Charleston: Arcadio.

Portland Investment Initiative website, 2018. http://wearepii.com/.

Sheldon Shafer, 2017a. "270-unit NuLu Apartment Complex Set to Break Ground in Fall." *Courier-Journal,* May 12.

Sheldon Shafer, 2017b "Brewpub or Distillery? Main Street Buildings Sold to Investment Group Led by Gill Holland." *Courier-Journal*, September 8.

University of Louisville, 2018. "Portland MFA Studio." https://louisville.edu/art/facilities-resources/portland-studio.

Nina Walfoort, 1999. "Armstrong to Push Downtown Housing." *Courier-Journal*, January 18, A1.

Part III
Varieties of boburbs

8   Crescent Hill

The bourgeois boburb

Louisville's other boburb is Crescent Hill – with a little more emphasis on the "bourgeois" part of "bourgeois bohemian." Crescent Hill was born suburban, and never really had a bohemian phase.

Rich families built country estates out along Frankfort Pike when it was the main route from Louisville to the state capital. In the 1840s the railroad to Frankfort was built beside the pike. This, in turn, made the area a convenient spot for the annual agricultural fairs of the 1850s.

What really made Crescent Hill part of the city, and gave it a name, was the construction of the main city reservoir at the top of the hill, some 33 feet above the riverfront, in 1879. This site was an immediate park-like attraction for visitors, and is still an important part of Louisville's infrastructure (Findling, 2012, 91).

With the completion of the water plant, it made sense to run mule-drawn streetcars out to Crescent Hill, Louisville's first "mule-car suburb." The railroad also had several Crescent Hill stations. In the 1880s a "Suburban Club" of matrons would regularly take the train into town to shop (Rogers, 1955, 32). By 1910 the streetcars were electrified, Frankfort Pike, now Frankfort Avenue, was paved, a public school created, and four Protestant denominations planted. Indeed, schools and (Protestant) Sunday schools were the first institutions built in Crescent Hill, and in many suburbs (English, 1972, 135). This is the beginning of true suburbanization.

In the 1890s Louisville began annexing Crescent Hill, which Crescent Hill resisted. This is a drama repeated over and over in the history of suburbs. The suburbanites fled the city for more freedom from regulation and taxes. They felt superior to the city, which they said was dirty, crime-ridden, and full of undesirable people. Yet the annexation effort was successful in Crescent Hill, as in all the streetcar suburbs, because the city also brought improved services, especially in sanitation and water control. Annexation turned Crescent Hill into a neighborhood of Louisville.

In 1908 developer Clarence Gardiner built in – and built up – Crescent Hill. Crescent Hill, he contends,

> is our only suburban district, and will remain suburban. Crescent Hill …
> is laid out on the village plan, wide streets and big yards, … until the

district has taken on a character so thoroughly suburban that no amount of increased population can ever change the suburban atmosphere of the place, and with the increased demand for room ... that comes with education in the better things of life, Crescent Hill will continue to grow in popularity and value, for it is the only suburb of to day that is the city of tomorrow ... the family seeking the joys of the country and the conveniences of the city has nowhere else to go.

(Quoted in Crescent Hill Improvement Club, 1978; ellipses in original)

Gardiner perfectly expresses the suburban dream – the joys of the country and the conveniences of the city – then and in the many waves of suburbs since.

The country edge of these suburbs remained attractive for charitable campuses. The Kentucky School for the Blind had moved from downtown to Clifton, adjacent to Crescent Hill, in the 1850s, when that was the edge of town. St. Joseph's Orphanage moved from downtown to Crescent Hill in the 1880s to what had been the site of the earlier agricultural fairs. The Masonic Widows and Orphans Home moved from downtown to the eastern end of Crescent Hill in the 1920s. The already-existing Sacred Heart Academy, built on the Lexington Pike, which paralleled the Frankfort Pike, was connected to the emerging Crescent Hill by new residential streets. In the 1920s the Louisville cultural institution of widest impact throughout the South, the Southern Baptist Theological Seminary, moved out to a property near Sacred Heart. In 1926 another land-intensive suburban amenity was built, the public golf course (Findling, 2012, 61).

Housing in Crescent Hill was built up in a fashion quite different from older Louisville neighborhoods. Since the first houses were country estates, there were lanes leading from Frankfort Avenue or the other country pikes back to large houses. As Crescent Hill developed as a suburban neighborhood, those lanes were then filled in with many, somewhat smaller, houses. The side streets connecting these lanes then drew smaller houses still, especially as they were a longer walk from the streetcar line. Frankfort Avenue then grew a long row of small shops along its southern side only, as the northern side of the street was occupied by the railroad tracks (English, 1972, 78).

Crescent Hill excluded African Americans almost entirely until after the 1960s. Unlike downtown areas, Crescent Hill never had significant industries that employed large numbers of black people. Likewise, its suburban development was well after the end of slavery, but during the era of racially restrictive covenants. Even today it is more that 90% white, and only 3% black. There are no Jewish institutions in Crescent Hill, though the city's first Jewish mayor lived there. Catholic institutions were also kept to the outskirts. St. Joseph's Orphanage and Sacred Heart Academy were in Crescent Hill, but only on the edges. The first Catholic Church was not built in the neighborhood until 1930, and even then was on the far eastern edge.

When the car suburbs began growing further east, Crescent Hill declined. While always a decent place to live and raise children, its base suffered after

World War II (Thomas and Thomas, 2011, 275). When the next town to the east on Frankfort Avenue, St. Matthews, built the first shopping mall in the region in 1962, the little shops on Frankfort Avenue suffered. Apartment buildings were starting to replace single-family homes around the edges.

Then, in 1974, a tornado hit. The very powerful F5 tornado walked across much of eastern Louisville. Two thousand trees in Cherokee Park, on the south side of Crescent Hill, were knocked down. In Crescent Hill itself, whole streets of houses were decimated.

The tornado marked an epoch in the culture of Crescent Hill. The churches came together to form Crescent Hill United Ministries to help neighbors recover, then kept going as a united services organization. The Crescent Hill Neighborhood Association, a loose advocacy group formed only a few years before, became a major advocate for rebuilding and re-imagining Crescent Hill. In 1982 much of Crescent Hill's core was placed on the National Register of Historic Places, a tool used by other boburbs to preserve the variety of old houses and stores (Thomas and Thomas, 2011, 283).

In 1988, in order to attract a restaurant that wanted to move from the Cherokee Triangle to Frankfort Avenue, the council voted to "go wet," allowing alcohol sales in the precinct. "Suddenly," concluded the most thorough local historian, "Frankfort Avenue was the place to meet and eat." The city spurred commercial development with small business renovation loans. By the mid 1990s, the newspaper called Frankfort Avenue in Crescent Hill "Renaissance Road" (Thomas and Thomas, 2011, 284). Jerry Abramson, Louisville's "mayor for life" from his first election in 1986 to becoming Lieutenant Governor in 2011, lived in Crescent Hill. He said of it:

> there's great camaraderie among the folks who live in the neighborhood from Halloween parties, block parties, to watching after each others' kids. … It's got Frankfort Avenue as a wonderful focal point and with lots of neighborhood restaurants and stores, and it doesn't have one fast food restaurant on Frankfort Avenue, not one.
>
> (Quoted in Thomas and Thomas, 2011, 287)

The intense pride in the local place, as opposed to the chain-store space, is a mark of a vibrant boburb.

Crescent Hill has the raw materials to be a boburb, as a walkable, leafy, dense settlement with a well-developed local retail spine. The Blue Dog Bakery recruited Carmichael's Bookstore, which in turn drew Heine Brothers Coffee to Frankfort Avenue. Crescent Hill has taken off as a kind of boburb, though, largely as a spillover from the Highlands. People looking to open independent stores or find affordable housing in a mixed-use neighborhood, who would have settled in the Highlands in the 1990s, started to see similar, but cheaper, places to develop in Crescent Hill in the 21st century. In 2010, *Southern Living* magazine named Crescent Hill one of ten of "the South's Best Comeback Neighborhoods" (Perry, 2010).

One Centre sociology graduate, trained in the language of the boburb, put it this way: "You move to the Highlands when you first get out of college, in your pot-smoking years. When you get married and settle down, but still want city culture, you move to Crescent Hill."

Indeed, many of the long-time residents of Crescent Hill had lived in the Highlands, or Old Louisville, or neighboring Clifton, in their youth. When they married, or when they had kids, they moved to family-friendly Crescent Hill. Most had looked at the Highlands, but couldn't afford a family house, or were daunted by the greater congestion. Several looked at the suburbs, but found them not walkable enough, not diverse enough, not cosmopolitan enough, and too bland. Several were drawn to Crescent Hill because it was more liberal than the suburbs. All appreciated the rich community interaction they found in their immediate neighborhood.

One couple fully embrace the walkable, liberal feel of Crescent Hill. When they were young parents in the 1980s, though, they "did not feel funky enough" for the Highlands. She wanted an old house with character because that is what she grew up with; he wanted a house with character in a close neighborhood because that is what he did *not* grow up with. They appreciated the short commute to work downtown. They were dedicated to public schools, and researched carefully which was the best option for each of their children. When their daughter married a suburban man, the young couple bought a big house in the suburbs "to get more house for the money." However, when the reality of raising a child came to them, they realized that a smaller house with more community was well worth trading for. They moved three houses down from her parents in Crescent Hill.

A man with long experience in the design business lived in Clifton, both as a bachelor and then with his wife when first married. His parents thought his wife would be raped and killed in this "inner city" neighborhood. In fact, though, they had no burglaries or any other crimes. He thought suburbanites fear the inner city because of the media, which only shows "the crazy things" that happen in the city. They moved to Crescent Hill to have more room for kids. They loved their neighborhood as affordable, close to his downtown office, walkable, and with diverse housing types. From his perspective as a designer, suburban housing was too bland, monotonous, and harder to customize.

Another longtime resident was an Ivy League corporate attorney, married, with kids – seemingly natural suburban material. Yet he wanted to be involved in his community, use public transportation, and support public schools. He reacted against huge McMansions and the kinds of neighborhoods that go with them. After an Old Louisville bachelorhood, he and his wife moved to Crescent Hill to raise their children. He became active in its religious ministries and its local government. He appreciates the sometimes-hidden class diversity of the neighborhood – he knows, though few of his neighbors do, that the neighboring apartments are subsidized Section 8 housing. He revels in the sheer number of different people one can encounter in Crescent Hill. He is proud to say that the people in his neighborhood like to talk about it, and are very protective of it.

Another older couple had lived in the car suburbs to raise their young children, but found they had no real contact with their neighbors. When they were ready to downsize, they realized they were blessed with a son who was a good handyman. This emboldened them to buy a fixer-upper in Crescent Hill, which they and their son gradually renovated. They love the walkable, dense neighborhood. Their daughter and her family eventually moved down the street, and both households are active in all the neighborliness of the block. An academic, he cited Jane Jacobs' evocative portrait of the rich community of Greenwich Village in the 1960s, famously described in *The Death and Life of the Great American Cities* (Jacobs, 1961) as the kind of thing they were looking for in their neighborhood.

Young couples with babies show the same enthusiasm for the neighborhood today as these older couples do. A liberal couple who work for non-profits appreciate the density of Crescent Hill, but wanted even more – when they could afford a house, they bought in Germantown. A politically moderate couple with a business career started dating on a college course in London, which made them want to live in a city. They, too, like the walkability, the community, the variety of things to do. They were impressed that their neighbors were "proud to live on this street."

A third couple with a baby and plans for more, active in a conservative church, were just as committed to Crescent Hill. He liked it because he had grown up there, and wanted to recreate the dense community he knew; she liked it because she had grown up on a farm, and wanted to be part of a dense community she had never known. While her family feared for them living in the city, he had enough urban experience to say "I rationally chose not to be afraid." They chose to be less interested in material accumulation than some of their college classmates — they contend that "it's cooler to have less stuff and be in the area where you can walk around." She offered an excellent summary of Crescent Hill as the more bourgeois of Louisville's boburbs: "it is a fusion of a family-friendly neighborhood and serves as a gateway to a lifestyle on the go."

References

Crescent Hill Improvement Club, 1978 (reprint). *Beautiful Crescent Hill, Illustrated,* originally 1908. Louisville: Bush-Krebs Company.

Judith Hart English, 1972. "Louisville's Nineteenth Century Suburban Growth: Parkland, Crescent Hill, Cherokee Triangle, Beechmont and Highland Park." Unpublished MA thesis, Division of Humanities, University of Louisville.

John Findling, 2012. *Louisville's Crescent Hill.* Charleston: Arcadia Publishing.

Jane Jacobs, 1961. *The Death and Life of Great American Cities.* New York: Vintage Books (Random House).

Rex Perry, 2010. "The South's Best Comeback Neighborhoods." *Southern Living*. https://www.southernliving.com/home-garden/best-neighborhoods#best-neighborhoods-crescent-hill-louisville.

John Rogers, 1955. *The Story of Louisville's Neighborhoods*. Collected stories from the *Louisville Times* of May 1955. Published by the *Courier-Journal* and the *Louisville Times*.

Samuel W. Thomas and Deborah M. Thomas, 2011. *Crescent Hill: Its History and Resurgence*. Louisville: SWT.

9 Norton Commons

The boburb in the fields

In the far northeastern corner of Jefferson County is a master planned community unlike all the fancy subdivisions built around it. Norton Commons was an attempt to bring the Cherokee Triangle out to the far suburbs. What was a big gamble when it began has been a commercial and social success. Those who like it tend to love it. On the other hand, quite a few people in the actual boburbs tend to regard Norton Commons, while a game effort, as artificial – "Stepford" and "Disney" are terms they often use.

The WAVE farm was an unusual venture. Founded by Louisville television pioneer George Norton, the farm was the site of agricultural experiments which were shown on WAVE television station in the 1950s and 1960s. When suburban development started to reach past the Snyder in the 1980s the Norton heirs wanted to develop the property. At the same time, though, they did not want it to be just another luxury subdivision, like the nearby Lake Forest. The owners were intrigued, therefore, when they were approached by developers with a different idea, to build a complete mixed-use "new urbanist" community, for a range of ages and social classes (Kaufman, 2014; Norton Commons, 2018a).

"New Urbanism" is a design movement to create walkable neighborhoods containing a wide range of housing and job types. It has developed a strong interest in environmental sustainability, as well. The movement looks to Jane Jacobs as its patron saint. The Norton Commons developers went to the leading new urbanist firm, Duany Plater-Zyberk, creators of Seaside, Florida, to make their dream community (Congress for the New Urbanism, 2018; Duany Plater-Zyberk, 2018; Jacobs, 1961).

Andrés Duany came to Louisville in the 1990s to study the Cherokee Triangle, as well as other functioning, walkable, mixed-use communities. In 2004 the developers broke ground on Norton Commons in a 600-acre territory in the crook of the Snyder and I71, 13 miles from downtown Louisville. At the heart of the community is a town square, with small commercial buildings topped by condominiums, around which radiate blocks of houses in walking distance. The houses have front porches, are close to the sidewalk, and smaller than the McMansions that other developers had found to be the most profitable. The houses have different designs on purpose, for aesthetic variety and to serve a range of tastes, pocketbooks, and household configurations.

Norton Commons is modeled on urban neighborhoods, but is still a suburb. It is sold to families with at least middle-class money. The vision statement on the website sells the walkable, mixed-use design as aimed at enabling parenthood and consumption:

> Our unique neighborhood offers your family the opportunity to do less planning and more living. With restaurants and retail shops just around the corner and plenty of green space for you and the kids, there's more to do, more to enjoy, and more life to be lived! Norton Commons isn't just a neighborhood – it's a lifestyle experience. Beautiful residential homes flow seamlessly into a charming town center, all connected by parks, walkways, pools and plenty of other upscale amenities.
>
> (Norton Commons, 2018a)

Unlike most suburban subdivisions, though, new urbanist communities intentionally aim at a range of ages and life stages, in both the housing types they embrace and the businesses and amenities they include (Besel and Nur, 2013).

The obvious exclusion entailed by Norton Commons is of people who cannot afford to live there. Today it has among the highest cost per square foot of housing in Louisville. To combat this, the designers intentionally made the houses and the lots smaller than the McMansion subdivisions. They included condominiums above the stores. They built apartments on the still-walkable edge of the neighborhood for both first-time householders and older people downsizing. The green and walkable parts of the community are free to use. Still, Norton Commons is not cheap to build in. And implicitly it rests on incomes earned elsewhere, either in other suburbs or all the way into Louisville.

The class exclusion element came to the fore in a fight over constructing a small 21-unit apartment building near the heart of Norton Commons as affordable housing, with upper limits on the incomes of residents in it. A small group of residents opposed the project. They argued that poor people would not be able to afford to live there, that public transportation to Norton Commons was not good, and that poorer residents would feel alienated. Another opponent, who owned a business in the neighborhood as well as residing there, said the apartments would lower nearby property values. "I earned the right to be out here, and now suddenly somebody is going to be put out here because they are what they are," he said, characterizing affordable housing as a "freebie." By contrast, more liberal residents supported the project, saying that diversity and inclusion were part of the new urbanist vision (Ryan, 2016). When the more affordable apartments were built, the opponents moved away, and the controversy disappeared.

There is idealism in the whole Norton Commons project, which is shared beyond the owners and developers. The Jefferson County Property Valuation Administration, the government agency that sets property tax rates, addresses the issue of diversity in the new urbanist community, compared to traditional suburbs:

What many may consider an exclusive planned suburbia revival is in fact a culturally, religiously and even economically diverse and accessible community. Owned and operated not by a corporate conglomerate but by a tight knit local family team, Norton Commons is a mecca of all things local. There are currently fifty-six local businesses including a plethora of restaurants, shops, doctor's offices, a fire department, a church, three different schools and a YMCA.

(Jefferson County Property Valuation Administration, 2018)

An experienced manager of high-end subdivisions said Norton Commons was different from other suburbs, and fosters more community, because there is a place to meet the neighbors.

The residents who like Norton Commons tend to love it. A dedicated "Highlands girl" and her husband went to a Norton Commons open house on a whim, which made them ask themselves "why buy a fixer-upper in the Highlands when we can buy new here?" A gregarious couple, they knew that, without children to connect them to the community, they would have to be joiners. They found there were many community-making groups to join. After they made friends with their porch-sitting neighbors, she joined the social committee to prepare mixers. The social committee sponsor monthly "Sips at Sundown," hosted in a resident's house, in which as many as 40 neighbors gather to drink and meet. Another advantage they noted compared to the high-end gated communities around them: their annual HOA dues of $800 were about a quarter of what they would pay in a residential-only subdivision, because the fees paid by the businesses offset much of the community maintenance costs.

One enthusiastic mother summed up her family's feelings about the community.

My sons best summed up living in Norton Commons ... "It's like living on vacation!" There are always people walking, running, walking their dogs, sitting on porches, playing in the parks, etc. and everyone is so friendly! I think we meet someone new every week! the neighborhood provides lots of opportunities for meeting people and socializing through weekly street parties in the summer, mostly concerts at the amphitheater, movie nights the lawn, parades & holiday parties, and a lot of different groups to join depending on your interests. We have a very active Mom's Group, singles groups, bourbon group, wine club, walking group, seniors, etc. Although my boys say it's like living on vacation, I often liken it to a more mature college experience!

She added an element of cosmopolitanism that I did not hear in other Louisville suburbs: "This neighborhood is full of transplants like me so no one here asks where I went to school (the infamous Louisville high school question) or which country club I belong to." All of this sounds like a true boburb.

The leading opponent of affordable housing in Norton Commons said she was drawn to the neighborhood because it reminded her of Disney. Those who dislike Norton Commons also draw the comparison to Disney, but not in a positive way. One columnist for the *LEO,* the Louisville Eccentric Observer, opened her comments on the affordable housing fight in Norton Commons this way:

> I grew up in the suburbs. As an adult, I lived as far from the 'burbs as possible. The separatist saga over "affordable housing" in Norton Commons intrigues me primarily because it smacks of classism, but also because I cannot imagine the desire to live in a sanitized, faux Highlands. Ew.
>
> (Houston, 2016)

Another woman, who had lived near Norton Commons before choosing Clifton, concluded "Norton Commons is bizarre, a Stepford world, insular. You aren't sure if you are allowed in the stores. Not an immediate sense of belonging."

Artists who lived in the Highlands were particularly skeptical of the Norton Commons experiment. They wondered why the kind of people who would like to live in the Highlands would buy this "very suburban, cleaned up" version. Another compared it to Fourth Street Live!, Louisville's attempt to create an attractive downtown entertainment district – the third attempt on that site. He thought, like Fourth Street Live!, that Norton Commons was an artificial attempt to make something alive (or Live!). A third echoed this theme of artificiality. "They're trying to manufacture a porch society, whereas what's odd and crazy about this society [in the Highlands] was organic," she said. Maybe, she offered, in 50 years Norton Commons might start to feel soulful.

At first, the whole concept of Norton Commons was treated as a risky gamble. The stores and houses were built on speculation. Soon, though, there was enough of a market for people who wanted to live in an urbanist, but not actually urban, community. The Norton Commons Trust, which owns the land, could sell lots to people who built their own houses, rather than building houses on spec. This, in turn, made the neighborhood more architecturally diverse and distinctive. Building a house to your own design also helps make a *space* into a *place* that people love and commit to. "'Norton Commons is thrilled to continue to engage in one-of-a-kind placemaking,' said Managing Director Charles Osborn III. 'We're all about walkable, mixed-usage, with an eye toward the future, but never forgetting Louisville's unique roots and history'" (Norton Commons, 2018b). In general, people who like the suburbs are not committed to any particular subdivision. Norton Commons may be the exception.

References

Karl Besel and Yusuf Nur, 2013. "A Comparative Study of Entrepreneurship in New Urbanist Communities." *Journal of the Indiana Academy of Social Sciences*, 16, 2 (Fall–Fall–Winter): 35–44.

Congress for the New Urbanism website, 2018. https://www.cnu.org/who-we-are/movement.

Duany Plater-Zyberk website, 2018. https://www.dpz.com/.

Holly Houston, 2016. "On Norton Commons and the Importance of Diversity." *LEO Weekly*, September 21.

Jane Jacobs, 1961. *The Death and Life of Great American Cities*. New York: Vintage Books (Random House).

Jefferson County Property Valuation Administration website, 2018. https://jeffersonpva.ky.gov/2015/08/26/so-what-is-norton-commons/.

Steve Kaufman, 2014. "Norton Commons Ten Years Old: Where Is It Now? Where Will It Go From Here?" *Insider Louisville*, June 26.

Norton Commons website, 2018a. http://www.nortoncommons.com/about-norton-commons/the-norton-commons-story/.

Norton Commons website, 2018b. "Norton Commons Releases Plans for North Village Amenities." April 18. http://www.nortoncommons.com/norton-commons-releases-plans-for-north-village-amenities/.

Jacob Ryan, 2016. "Fight Over Affordable Housing Erupts in Norton Commons." *WPFL*, September 15. https://wfpl.org/fight-over-affordable-housing-erupts-norton-commons/.

10 Russell and Shawnee

Imagining a multiracial boburb of the future

If there is to be a multiracial boburb in the near future in Louisville, the most likely place is Russell and Shawnee. These two adjacent neighborhoods start at downtown and run west to the Olmsted park on the river. Right now, when both neighborhoods are deep in the poor black ghetto, such a future may seem hard to imagine. But the great original sin of America – anti-black racism – is slowly being overcome, and Louisville has long had an important role in that transition. The process of boburbanization moves on, and will likely some day move to other parts of the city. Russell and Shawnee seem to me the most likely place for these two processes to come together. They have attracted a concerted redevelopment effort, they are adjacent to Portland, and they have the people to make a culturally and economically rich neighborhood on an African-American base. Russell and Shawnee have the bones of a dense, walkable, streetcar neighborhood with historic bourgeois buildings begging for restoration.

Shawnee is where you get if you head due west out of downtown Louisville until you stopped by the Ohio River. Just west of Portland, the river, which flows east to west across most of Louisville, makes a right-angle turn southward. This corner creates the natural western boundary of Louisville, and has been a brake on urbanization. Whereas the parts of southern Indiana just north of downtown – Jeffersonville, Clarksville, and New Albany – have grown into small cities, the land west of Shawnee, just across the river, is still mostly farmland and forest.

At this western edge of Louisville the city built another Olmsted park, Shawnee, along the river, in 1892. Three main east–west streets ran out to the park – Market, Walnut, and Broadway, the latter marking the park's southern boundary. The streetcar line was extended to the park in 1895, and development quickly followed. The blocks nearest the park, in particular, were filled by middle- and upper-middle-class families, whose homes still remain. The commercial district serving the streetcar lines were built along the boulevards (English, 1972, 47). In 1927 a public golf course was added to the park. As was often the case with streetcar lines, there was a popular private amusement park, Fontaine Ferry Park, which operated from 1905 to 1969.

Shawnee was originally predominantly white. Russell, the "Harlem of Louisville," was the center of black life. However, the entire west end of Louisville south of Market Street (that is, south of white Portland) was increasingly the segregated black park of town. In 1923 the city opened a smaller park, Chickasaw, just down the river from Shawnee. Shawnee Park, and Fontaine Ferry Park, were reserved for white people, while Chickasaw – which had no equivalent private amusement park – was for African Americans. This segregation was an increasing source of grievance. Lawsuits begun in the 1950s finally succeeded in desegregating the public parks in 1963. However, a second round of race riots in 1969 damaged Fontaine Ferry Park, which closed. White flight, and middle-class black flight, which had begun in earnest after the Russell riots the previous year, accelerated in the early 1970s.

Today Russell and Shawnee are about 90% African American, in a city that is 22% black. Walnut Street, which runs through the heart of Shawnee, was renamed for Louisville's most famous African-American son, Muhammad Ali, in 1978 (Adams, 2016). Into the 21st century, locals who grew up there and went away to college tended to settle in other, more middle-class parts of Louisville. Whereas the proportion of adults in Jefferson County with a bachelor's degree is about 28%, the percent for Russell and Shawnee are in single digits (Open Data Network, 2016; Louisville/Jefferson County, 2018). The grand 1920s Shawnee High School had declined markedly by the end of the century. However, it was redeveloped as (of all things) an aerospace magnet high school, with the very 2010s new name of Academy@Shawnee. The school has begun a slow turnaround.

One important factor that may help push Russell and Shawnee toward revitalization as a boburb is that they are across Market Street from Portland. If the Portland experiment works to spur a boburb into being by planned action, it is almost inevitable that the movement would spill southward. Black Shawnee has a lower poverty rate, and a higher home ownership rate, than white Portland (Bourassa, et al., 2003). Shawnee contains some of the city's best examples of late-19th-century architecture, including the Italianate, Renaissance Revival and Victorian Gothic styles (Pillow, 2004). Shawnee has the park and the housing stock. With a 14% vacant property rate, these west end neighborhoods could absorb a sizable new population without displacing the current residents.

The biggest efforts at renewal, though, are being made in Russell (Vision Russell, 2018). The neighborhood is in easy walking distance of downtown and of the river. A Russell-based redevelopment project, like the one in adjacent Portland, has been developing the economic corridor on Broadway running east and west, and on 18th Street, running north and south (OneWest, 2018). The Kentucky Center for African American Heritage has been constructed in a renovated factory (Kentucky Center for African American Heritage, 2018). The scary public housing project is being replaced by more mixed, and better regulated, housing. An ambitious public/private venture to create a "FoodPort" in Russell collapsed in 2016. Like many poor neighborhoods, Shawnee is a "food desert," with no full-scale grocery store. The food port also included food production, centered on a larger vertical farm (Downs, 2016). While this effort did not succeed, others clearly see not only the need, but the possibilities, of innovative economic development in the west end. Instead, a track and field training center will be developed in that space, building on black Louisville's strong history in track events (Finley, 2017). Perhaps the track team could be a unifier for the west end, as the swim team is for the suburbs.

There are, of course, obstacles. The first is that the west end, especially the black west end, is feared as a place of crime and violence. A neighborhood association activist in Shawnee was the only person in all of my interviews who mentioned the major in charge of the local police district as an important person in the life and future of the neighborhood. Racial fears have made most white people, especially in the suburbs, afraid of going "west of Ninth Street," ever, for any reason. Louisville's black ghettos do, indeed, have a higher crime rate than other parts of the city. However, suburbanites rarely differentiate the various different neighborhoods that make up the west end. The issue is not simply racial fear, though – black college graduates don't move to the west end, either, even if they grew up there. They may come back for church on Sundays, but they raise their own children in the safer parts of town.

The exciting new possibility of Shawnee as a boburb is of a truly multiracial and multicultural knowledge-class neighborhood. Boburbs by their nature yearn for diversity and inclusion. But, thus far in their historical development, they have been predominantly white, in Louisville and elsewhere. As the United States becomes less white (as that category is currently understood), it seems almost certain that all kinds of middle-income neighborhoods in the future will be more mixed racially, ethnically, and culturally. The places that figure out how to make such a place early and intentionally may have an advantage in the emerging culture and economy of the future. In Louisville, Shawnee and Russell seem the most likely foundation on which to build.

References

Lauren Adams, 2016. "Daughter Remembers Father's Push to Have Street Renamed Muhammad Ali Boulevard." WLKY television news website. June 8. http://www.wlky.com/article/daughter-remembers-father-s-push-to-have-street-renamed-muhammad-ali-blvd/3767744.

Steven Bourassa, Eric Schneider, Bruce Gale, Jack Trawick, 2003. *Housing Conditions and Challenges in Louisville's Western and Central Neighborhoods: A Report to the Louisville Community.* Unpublished report sponsored by the Neighborhood Reinvestment Corporation, Fannie Mae, Freddie Mac, and Enterprise Foundation.

Jere Downs, 2016. "West Louisville Food Port Cancelled." *Courier-Journal*, August 18.

Judith Hart English, 1972. "Louisville's Nineteenth Century Suburban Growth: Parkland, Crescent Hill, Cherokee Triangle, Beechmont and Highland Park." Unpublished MA thesis, Division of Humanities, University of Louisville.

Marty Finley, 2017. "$30 Million Track and Field Facility to be Developed in the West End." *Louisville Business First*, September 19.

Kentucky Center for African American Heritage website, 2018. https://kcaah.org/.

Louisville/Jefferson County website, 2018. "West Louisville Strategies for Success." https://louisville.edu/cepm/westlou/west-louisville-general/vacant-properties-campaign-presentation/.

OneWest website, 2018. https://onewest.org/.

Open Data Network, 2016. *Louisville/Jefferson County Graduation Rates.* https://www.opendatanetwork.com/entity/310M200US31140/Louisville_Jefferson_County_Metro_Area_KY_IN/education.graduation_rates.percent_bachelors_degree_or_higher?year=2016.

John C. Pillow, 2004. "Shawnee: Farms of the 1800s Gave Way to Park and Dignified Homes; Racial Makeup Has Changed." http://orig.courier-journal.com/reweb/community/placetime/city-shawnee.html.

Vision Russell website, 2018. https://visionrussell.org/.

11 Boburbs elsewhere

The boburb as an ideal type of a neighborhood is found nowhere in perfect form, but different kinds of boburbs can be found in many cities with variations on the theme. Part of the point of a boburb is that it fosters distinctive culture, so they should be a bit different from one another. Moreover, the form we see in Louisville, built along a few of the old streetcar lines, is reproduced in some cities; in others, though, boburbs arose with a different history. And it is the nature of any city to be a flowing ecology. The special combination of features which produce a boburb now were not quite there before, and will likely not always remain so. The boburb may not be as ephemeral as a bohemia, but it is not forever.

This chapter will look at five specific neighborhoods around the United States with some claim to be boburbs today:

- Travis Heights, Austin
- Wicker Park, Chicago
- Fishtown, Philadelphia
- Virginia Highland, Atlanta
- Pearl District, Portland

Of all the established and potential boburbs in the United States, I want to lift up these few because they are, in one way or another, already famous. Bill Bishop begins *The Big Sort* with an account of how he and his wife selected the Travis Heights section of Austin on the vibe they got while driving through it, before they knew anything about Austin neighborhoods. Richard Lloyd focused on Wicker Park as it was becoming *Neo-Bohemia* in the 1990s. Charles Murray tagged Fishtown in Philadelphia as the ideal-typical white working-class neighborhood in *Coming Apart* – only to admit at chapter's end that it was being taken over by hipsters. Virginia Highland in Atlanta is a streetcar suburb turned cool. The Pearl District in Portland is the heart of the exemplary hipness parodied – and partly celebrated – in the television show *Portlandia*. I commissioned a new survey of each neighborhood, which gives us a sample of 50 residents in each place. They will give us a window into the experience of each neighborhood for both bobos and those they live among.

Three of these cities – Austin, Portland, and Louisville – have fostered notable "Keep [city X] Weird" campaigns. This has been the template for other campaigns to preserve distinctive local culture against routinization. In several cities the local Independent Business Association has adopted the slogan, or a variant on it, to mobilize consumers to see local businesses as an asset to a distinctive community, against the chain stores. This commercialization of the message is seen by some as a cooptation of what was originally an anti-commercial idea. Still, one of the distinctive features of a boburb is that it mixes in commerce with residences and non-profits.

"Keep Austin Weird" was originally suggested by a listener and supporter during a public radio fund drive. The radio station liked the idea and promoted it. The campaign really took off, though, when an independent bookstore printed hundreds of bumper stickers with the slogan as part of their fight against a Borders chain bookstore. There is some irony that the original "keep our place weird" campaign was waged by one bookstore against another – that is, by a central knowledge class institution against another. While the phrase refers to the whole city, it was most popular in just a few Austin neighborhoods, which saw themselves as the repository of the "weirdness" and independence of the whole city. A further irony is that this phrase, which was created for a public radio station, was copyrighted by an opportunistic tee-shirt printer, who was then sued by the bookstore that made it famous (Long, 2010, 94).

"Keep Portland Weird" has become so well known that many think the phrase, and the movement, originated there. In fact, this is an illustration of cultural diffusion among independent knowledge class networks. A Portland bookstore owner knew the Austin bookstore owner, thought this defensive campaign would be useful at home, and copied the idea. Originally meant to promote independent businesses, as in Austin, it has grown to become the unofficial slogan of the city (Wikiwand, 2018).

"Keep Louisville Weird" is directly based on the Austin and Portland campaigns. It is the slogan of the Louisville Independent Business Alliance. The campaign began in the Highlands in the mid-2000s. It is sometimes expressed as "Keep the Highlands Weird." Ear X-tacy record store and Heine Brothers Coffee Company have been notable proponents of the theme. They distributed the "Keep Louisville Weird" bumper stickers well beyond their own clientele. The Highlands features a mural of Louisville native Hunter S. Thompson. His famous quote, "When the going gets weird, the weird turn pro," is often cited as a motto of commercialized uniqueness.

Travis Heights, Austin

When journalist Bill Bishop and his wife drove into Austin, Texas, to start a new life, they knew nothing of Austin neighborhoods. The Bishops had lived in the Highlands in Louisville, so knew what kind of neighborhood they liked. "We didn't have a list of necessities … as much as we had a mental image of the place we belonged," he says of their search strategy. "We didn't intend to move into a community filled with Democrats," he concludes, "but that's what we did – effortlessly, and without a trace of understanding about what we were doing." He describes their new home, Travis Heights, thus: "a shady neighborhood of dog walkers, Jane Jacobs-approved front porches, bright paint, bowling-ball yard art, and YOU KEEP BELIEVING, WE'LL KEEP EVOLVING bumper stickers."

Travis Heights is the South Austin neighborhood bounded by, and on the hills above, Congress Avenue, the city's famous bar, restaurant, and music-venue street. There were streetcars down Congress at the beginning of the 20th century – the street itself was not paved until 1905 – but they were all gone by World War II. Congress was one of the few streets that bridged the Colorado River, now dammed to create Lady Bird Lake, which forms the northern boundary of the neighborhood. Congress was a busy commercial street until the 1960s, when the opening of Interstate 35 took much through traffic away. I35 forms the eastern boundary of the neighborhood.

In the 1970s and 1980s cheap rents drew eccentric stores to South Congress Avenue, and artists and musicians to the neighboring residences. The independent music scene, in particular, became famous, centering on the Armadillo World Headquarters concert hall, which opened in 1970. Today, the SoCo (South Congress) neighborhood is as famous for its eclectic food offerings as it is for music and art.

Travis Heights and adjacent Fairview Park were developed in the 1920s as an upper-middle-class neighborhood, though the actual housing stock varied quite a bit. The residences range from large Victorians on the highest points, with views of the state capitol building and downtown Austin, to more modest bungalows set among the trees. There are also some apartment buildings, especially along the commercial streets. This mix of housing stock has helped Travis Heights attract families from a wider class range, and broader life-cycle moments, than the average car suburb can.

Bill Bishop's point in *The Big Sort* is that neighborhoods like Travis Heights have become increasingly polarized by politics against their antitheses, the gated communities and exurbs that ring Austin. Travis Heights was attractive to the Bishops because it supported their liberal politics. The same culture that his family found so attractive was also somewhat intolerant of conservative voices. He recounts a story of the lone vocal conservative on the neighborhood chat forum being silenced, and eventually moving to the suburbs. The suburbs are also politically polarized the other way, but they are more likely to avoid conflict by simply not discussing politics. A boburb, by contrast, is interesting to its residents *because* of the vibrant and (somewhat) varied expression. It goes with the bright paint (Bishop, 2008).

I asked a sample of residents of each neighborhood "Of all the reasons that led you to choose where you live now, what were the top two or three?" A very liberal retired Latina school teacher, married to a postman, who raised two children in Travis Heights wrote "When I moved to this neighborhood, no one really wanted to live here. Over the years, it's become the place where everyone wants to live!" A moderately conservative white salesman, married to a speech pathologist, who had also raised two children there, chose Travis Heights to "live in a city versus suburbs, be able to walk to restaurants and entertainment, vibrant." A childless white Millennial who lists her occupation as fashion e-commerce and her partner's as private equity chose it for "location and ease!" The younger, richer cohort may show how the neighborhood is maturing as a boburb. A white widower who had raised three children there praised Travis Heights as safe, clean, and with an easy walk to the library, though he worried about children crossing busy streets. A married Latino teacher with three children thought the main benefit of raising children there were their friendships.

Austin is the home ground of the "Keep Austin Weird" campaign. The core proponent now is the Austin Independent Business Alliance, which is headquartered just outside of Travis Heights. The phrase embraces all of Austin, but the heart of the support is in the boburbs. While anti-market bohemians may embrace "weirdness," it is more in the line of bourgeois bohemian independent commercial interests to make a slogan and a movement of it. The inventor of the phrase was a local college librarian, but it was an independent bookstore that made the campaign, and the bumper stickers, famous (Long, 2010). When I was describing the idea of a "boburb" to an officer of the Austin Independent Business Alliance, I mentioned that in the boburbs, as compared to the deed-restricted suburbs, there was always a purple house. She laughed. "In my neighborhood," she said, "*my* house is the purple house."

Wicker Park, Chicago

Richard Lloyd studied how the Wicker Park neighborhood of Chicago was transformed in the 1990s from a gritty slum to a bohemian quarter. Young artists flocked there because it was cheap and easily accessible by the elevated train from the School of the Art Institute and Columbia College, where many of them were, at least nominally, students. What had once been a Polish working-class neighborhood had lost its factories and most of its working class. By the 1970s it had decayed into a land of Mexican gangs and hookers. The grittiness of the neighborhood was why it was cheap. But, as Lloyd found, "grit is glamor" for bohemians. The artists saw themselves as in opposition to bourgeois, suburban, corporate culture. They took the life of the criminal underclass as more *authentic.* Their own participation in criminal life usually did not go past using drugs. Nonetheless, they saw value in the grit of their neighborhood for keeping the yuppies out, and as proof of their own adaptability, toughness, and tolerance for diversity (Lloyd, 2006).

The bohemians saw themselves as art producers. They liked to talk to other art producers about art and the making of it. The iconic place to have such conversations for three centuries has been the coffeehouse (Ellis, 2004). The coffeehouse is a central location for would-be artists to create the critical mass of community that turns an aggregation of individuals into a bohemia. The wonderfully named Urbus Orbis coffeehouse was that gathering place in Wicker Park at the crucial bohemianizing moment. The name evokes both the dense urban place that bohemians need and their cosmopolitan aspiration to affect the whole world. The bars were also places to congregate, but even more so, they were a way for artists to make money to support art production in their off hours. As Lloyd found, though, the bar workers and baristas were more effective at presenting themselves as works of art – which increased tips – than they were in actually producing art.

Art producers, though, need art consumers. The "yuppies" came to Wicker Park to participate in bohemia, to look at the artists, and to consume their work. They went to the bars to hear punk bands, like Veruca Salt, and see the art in storefront galleries. Soon enough, more bourgeois bohemians – bobos – started moving in to Wicker Park. They opened design firms and art distribution businesses to capitalize on the talent accumulating in the neighborhood. The bohemia was being transformed into boburbia.

The bohemians lamented that their bohemia was "over." Lloyd studied the history of bohemias, going back to the original bohemian quarter in Paris in the 1840s. While the Romantics of the early 19th century imagined their revolt against rationalization (and capitalism) as a rural phenomenon, the actual Romantic artists wanted to live and work in a dense community, with urban resources for artistic production. The perpetual conflict between anti-market "authenticity" and making a living means that bohemias are inherently unstable. If the artists fail to make a living, their bohemia falls apart. If they succeed in drawing a public, the yuppies, bobos, or whatever their equivalent is called, gentrify the neighborhood into something else. This is why Lloyd concludes that "The sense of being always already over is an apparently structural feature of both classic bohemias and their contemporary heirs" (Lloyd, 2006, 237).

A very liberal Millennial white woman in advertising, who lives with an illustrator, finds Wicker Park to be convenient to work and downtown, with lots of culture in walking distance, and relatively safe. A very liberal white lawyer, married to a video producer, likes the convenience and the liberalism of the neighborhood. A somewhat conservative married white couple, she in technical sales, he an engineer, like the area for its green space, proximity to the train and the highway, and the ease of walking. A young, politically moderate white man in retail, married to a social worker, liked the "nice/pretty/safe neighborhood," which has lots of cool places to eat around it, and is somewhat affordable. A moderate Baby Boomer white couple who had raised a child there praised the diversity and culture of Wicker Park for raising children, but worried about the increasing crime and bad public schools. A Latina single mother with two children found that the neighbors are like family and there are a lot of opportunities for kids. This rich mix of bobos, bohos, and working-class residents from an earlier period are fundamental to the appeal of the boburb.

Since *Neo-Bohemia* was published in 2006, the gentrification of Wicker Park has continued apace. Urbus Orbis closed in 1997, a mere nine years after it opened. The meteoric career of this central coffeehouse shows how the process of boburbanization makes bohemia "always already over." In the spring of 1989 the coffeehouse opened, the "first white business" in a pre-dominantly Mexican neighborhood. Within a year, it had become such a gathering place for the already existing aggregation of artists and musicians that other arts-adjacent businesses opened nearby. Within two years, the Chicago *Sun-Times* proclaimed that "Looking intellectual is in vogue at Urbus Orbis." The neighborhood arts festival, Around the Coyote, now one of the largest in the city, was born there. Urbus Orbis sponsored poetry readings and musical showcases from the beginning. It became such a hot venue for new musical talent that, in 1993 *Billboard* declared Wicker Park "Cutting Edge's New Capital." But by 1995, owner Tom Handey says, bohemia had peaked there. Rising rents drove out the artists, as well as the poor residents who had lived there before, replaced by people who wanted the lifestyle of bohemia with the comforts of the bourgeois life. "It seems to be part of the life-cycle of an artsy-fartsy neighborhood," Handey concluded. Two years later, the building's owner refused a new lease, thinking she could do better than a coffeehouse (Huebner, 1997).

Fishtown, Philadelphia

Fishtown, as the name suggests, is a river neighborhood, one of the oldest in the city. Penn's Landing Park, where William Penn first landed, is in Fish-town. It was a typical white working-class neighborhood for almost three centuries. Fishtown was so representative of a white working-class community that Charles Murray, in *Coming Apart*, used it as the name for his ideal type of a white working-class neighborhood. He analyzed the real Fishtown, and many other neighborhoods like it, in comparison with the other end of the white class structure, "Belmont." Murray saw increasing polarization between the two ways of living, with the Belmonts of America increasingly educated, married, well paid, and civically involved, while the Fishtowns were heading in the opposite direction.

And then, Murray allows,

> In the 2000s gentrification came to Fishtown. ... Fishtown had cheap housing compared to more fashionable neighborhoods, it was close to downtown Philadelphia, and it was reasonably safe. ... And so first the pioneers – the artists and musicians without much money – started to move into Fishtown. In the last few years, [he wrote in 2012] affluent young professionals have expanded their beachhead.
>
> (Murray, 2012, 223–224)

Bohemia quickly draws the boburbans.

A moderate white couple of teachers chose to raise their children in Fishtown because the area is close to their community of friends, and within walking distance to their favorite stores, favorite restaurants, and public transportation. A divorced restaurant owner likes the diversity of the neighborhood both for himself and as a benefit to his children. A woman who works for a university chose the neighborhood for its low rent (when she first moved there), easy access to restaurants, shops, grocery stores, and proximity to her other friends. Though she had no children herself, she thought a benefit of raising children there would be the diversity, nearby parks, other children living in the area, the several community programs, and active community members. A very liberal artistic professor, living with a graduate student, bought in to Fishtown when it was affordable and close to work. He is involved in the neighborhood association, a political party, as well as artistic, environmental, and social justice organizations. Though he does not have children, he thinks the neighborhood is kid friendly, with good magnet schools, childcare, activities, food, and a variety of other things.

Scholars who have studied the bohemianization of Fishtown note that the artists there do not have the traditional anti-bourgeois ideology that we expect from bohemians. These scholars think they have found a new bohemian subtype, which they call an "artistic bohemian lifestyle community." The selling of the bohemian lifestyle as, itself, an attractive commodity means these kinds of bohemians are collaborating with the Establishment, and are therefore complicit in their community's demise (Moss, Wildfeuer, and McIntosh, 2017). I think what these scholars have found, instead, are boburbans, who are making a different kind of community than a traditional bohemia.

A journalist's attempt to find out why Fishtown became "America's hottest new neighborhood" lifts up some features we have already seen in looking at boburbs, and some that are distinctive to that specific place. Fishtown was always a dense neighborhood, mixing housing, retail, and manufacturing firms of all kinds. When the factories shut down in the darkest days of the 1970s and 1980s, the residents still had pride in the neighborhood and defended it, informally and through organizations. Fishtown is close enough to center city that outside – but small-scale – developers spotted it as the Next Big Thing. What they did next was the crucial right step – the developers met with the locals, over and over, to make development (and class change) as shared a process as they could. The dive bar kept its name and basic look, but added craft beers and indie bands. The new apartments were built in scale with the existing houses. La Colombe, Philadelphia's leading local coffee chain, moved from tony Rittenhouse Square to build its roastery in an old Fishtown factory. Fishtown had long had trolleys, like other boburbs – but unlike many boburbs, Fishtown's trolleys were still running. Moreover, Fishtown was shaped by the El, the elevated train line that ran above the main commercial street straight to center city, nine minutes away. Indeed, Fishtown is a model for "transportation oriented development," a hot idea in city revitalization at the beginning of the twenty-first century (Taylor, 2018).

Virginia Highland, Atlanta

The neighborhood that grew up around the intersection of Virginia Avenue and Highland Avenue, about four miles northeast of downtown Atlanta, owes its existence to the streetcar. In 1890 the Nine Mile Trolley went out from downtown along Virginia, then turned back to toward town down Highland, making a nine-mile circuit. From that time into the 1920s bungalow neighborhoods sprang up in walking distance of the trolley line. Businesses grew near the Virginia-Highland intersection, including the trolley maintenance barns. The Olmsted firm had been commissioned to develop the Druid Hills neighborhood, just east of Virginia Highland. The leafy, walkable, curving streets of that fancy development affected the model followed in the more middle-class Virginia Highland. Parks, schools, and churches of the Protestant Establishment (Episcopalian, Presbyterian, and Congregationalist) soon followed.

After World War II the streetcar line failed. In the 1960s the neighborhood started to decline, as middle-class residents moved to the far suburbs. The proposal to run Atlanta's Perimeter highway, Interstate 285, through the neighborhood was the crisis that ultimately saved the neighborhood. In 1975 the Virginia-Highland Civic Association was founded to fight the highway. This effort, ultimately successful, created an organization and identity for the neighborhood. Young families returned and renovated the housing, new businesses developed, and the seeds of the boburb were planted (City of Atlanta, 1988).

By the 2000s, Virginia Highland had become a convivial small village within Atlanta, and a destination for bobos from all over to come and hang out. The bars, restaurants, and local shops made it a distinctive place. Inevitably, rents and house prices rose, and parking became a persistent problem (Kleine, 2001). By the 2010s Virginia Highland had matured into a fully walkable, mixed-use boburb. Local magazines awarded it "Best Walkable Neighborhood" (*Creative Loafing* Best of Atlanta 2012), "Best Overall Neighborhood" (*Creative Loafing* Best of Atlanta 2011), and "Favorite Overall Neighborhood" (*Atlanta Magazine*, June 2011).

The residents had lots of love for Virginia Highland. In my survey, 70% liked it a great deal. Among the reasons they gave:

- Ability to walk to restaurants and shops, interesting and diverse areas to walk to, lots of young people;
- Big house in fun cool neighborhood;
- In the city, conveniences, safety;
- Aesthetic, able to be car free, cultural diversity;
- Older neighborhood, people out walking, low traffic;
- Location to things to do and services. The beauty.

I asked about the benefits of raising children in Virginia Highland. Nearly all cited good schools in walking distance. One wrote "They can walk to coffee shops that are nearby their school and meet their friends. They will have more opportunities to hang out with their friends and play. There are two parks nearby where they can play." Several noted the diversity of kinds of people who live there.

The Pearl District, Portland, Oregon

The Pearl District was created as an industrial zone just north of downtown on the west side of the Willamette River. In the late 19th century it was built as a major rail center, serving Oregon's signature industry, timber processing. Worker housing was built amidst the industrial plants. Later, factories of all kinds were built, especially for furniture making. In the 1960s an economic recession and the general abandonment of cities for the suburbs decimated the Pearl District. In the 1970s and 1980s it was largely derelict, with many homeless residents. Artists started moving in for the cheap housing and the grit.

When the city and developers worked out a master plan to redevelop the whole river front, the Pearl became a central site of renewal and gentrification (in the good and bad senses). The city built a new streetcar line to connect the whole northwest side of the city. One of the district's most famous businesses, the enormous Powell's bookstore, helped attract other knowledge class businesses. An art college moved into the former central post office building. The armory was converted to a theater. A huge abandoned brewery became a warren of local stores. Art galleries and brewpubs opened. Coffeehouses proliferated. Midrise apartments were built at a rapid clip. There were no detached houses to begin with, and now housing is overwhelmingly in apartments and condominiums (Bruce Johnson, personal communication, October 20, 2018). As a long-time Portland architect and developer put it, "Earlier art galleries flocked to the district replacing the pioneering artists who had embraced the grittiness of the Northwest Industrial Triangle. The artists disappeared along with the grit" (Johnson, 2018). By the 2010s, *Forbes* declared it one of the top five "hippest hipster neighborhoods" in the United States (Brennan, 2012).

A professional dancer liked the Pearl District because it was close to her workplaces, was safe, and she liked the "quality of neighborhood." A single Latina "usher and artist" likes the area because it is walkable, the rent is affordable, and she feels safe. Quite a few residents were retirees downsizing to a low-maintenance building with easy transportation. A moderate Millennial white woman, working in finance and married to a chef, gave a more extensive answer than most about why they chose the Pearl District.

> Being in the city center so we would not need a car and be within walking distance of our jobs. Having amenities in the building. We needed a park nearby for our two dogs. More than anything I needed to live someplace that would inspire me to do and be more.

The Pearl was not very residential when the renewal began, so few families were displaced or saw their homeplace greatly changed. Most of the people who live there now, at its peak fashionability, have chosen it because it is a boburb.

The character of the boburbs

I commissioned Qualtrics Research Services to survey 50 respondents in each of the zip codes containing our five boburbs, for a total of 250 responses. We can compare this with my survey of 66 Highlands residents who are alumni of Centre College (surveyed September 2017). The results confirmed the general picture of a boburb: liberal, young, white but not wholly so, with a sizable single group, and an even bigger childless group. More women than men answered the survey, but that may reflect who answers surveys in general. Table 11.1 shows the percentages responding in the named way in each community.

We can compare this sample with figures for the whole neighborhood (Niche.com, 2018). We can see that the boburbs, for all their differences from one another, are each richer than the national average and much better educated than the national average (Table 11.2). They are also more racially diverse than are most middle-class and, especially, most upper-middle-class neighborhoods.

As expected, these neighborhoods are similar to one another, but not identical. Regional differences matters, especially for which non-white groups are most common. Virginia Highland (Atlanta), for example, is more African American, Travis Heights (Austin) more Latino/a, and the Pearl District (Portland) is more East Asian. Whether the neighborhood was more working class or more middle class to begin with still has an effect now. Fishtown and Wicker Park were built for blue-collar workers, whereas Virginia Highland

Table 11.1 Various boburb characteristics, compared to the Louisville Highlands.

All numbers as a % of that neighborhood's respondents	Travis Heights	Wicker Park	Fishtown	Virginia Highland	Pearl District	High-lands
Liberal (some-what + very)	58	52	40	54	62	65
Millennial or younger[1]	48	58	56	46	34	51
Female	64	68	72	52	60	54
White	64	70	86	86	84	100[2]
Never married	34	28	42	32	34	27
No kids	78	60	52	64	62	59

[1] Born after 1980
[2] The Centre College alumni are somewhat whiter than the Highlands as a whole.

Table 11.2 Various boburb characteristics compared to national characteristics

All numbers as a % of that neighborhood's respondents (national percentages in parentheses)	Travis Heights[1]	Wicker Park	Fishtown	Virginia Highland	Pearl District	High-lands
Household income $75,000–149,000 (13.37)[2]	26	32	29	27	27	25
Household income $150,000+ (11.32)	14	32	12	28	23	8
Bachelors degree (19)	41	46	32	38	40	27
Master degree + (12)	22	27	15	39	32	20
White (61.5)[3]	66	67	81	82	83	89
Black (12.3)	3	5	7	6	4	5
Latino/a (17.6)	27	19	5	5	4	4
Asian (5.3)	3	6	4	2	6	1

[1] Travis Heights was not listed separately. These figures are for the whole zip code.
[2] National household income figures from 2014 Census estimates.
[3] National racial figures from 2015 Census estimates. Other races and mixed race groups not included.

originally served white-collar streetcar commuters. Most importantly, I believe, these neighborhoods differ in how far along they are in the process of boburbanization now. Virginia Highland has been colonized by bourgeois bohemians for decades. Fishtown has only had bobo "new fish" among the working class "old fish" in the past few years.

The driving question of this whole study has been "why do people choose to live where they do?" In analyzing these surveys, therefore, the main "dependent variable" has been how much each resident liked their neighborhood, on a five-point scale from "Like a great deal" to "Dislike a great deal" (Table 11.3). Comparing the five boburbs around the country reveals more similarities than differences in what kind of person found them to be a treasured place, and who did not. Not surprisingly, many more residents like than dislike the place they have chosen to live. Comparing only the extreme positive opinion – like a great deal – with the total of both negative opinions – dislike somewhat plus dislike a great deal – still yielded more than three-to-one positive to negative. These were the percentages on each end (ignoring scores of "like somewhat" and "neither like nor dislike"):

Table 11.3 How the residents feel about their boburb

Neighborhood	Like a great deal	Dislike somewhat or a great deal
Travis Heights	56	8
Wicker Park	64	8
Fishtown	40	12
Virginia Highland	70	2
Pearl District	60	14
Highlands	94	0

Clearly, the experience in Fishtown today is more varied than in Virginia Highland. Perhaps this is because the dramatic transformation of Fishtown is so recent, whereas Virginia Highland has been boburbanizing for decades. Or maybe Southerners are just more polite than people in northern cities.

When we look in detail at what makes a difference between those who really like their boburb and those who don't, the first thing that stands out is what are *not* the main reasons. The sociologist's go-to explanations – race, class, and gender – do not seem to be the main drivers of these different experiences. Nor is generation a big differentiator, as I thought it might be in the most rapidly changing neighborhoods. Both the "like a great deal" and the "dislike" groups, in each neighborhood, tend to be predominantly white, educated, with more female respondents. Both groups also skew young and liberal. Most people in both groups are optimists, and most think their neighborhood is getting better (even the ones who don't like living there themselves).

What differentiates the happy residents from the unhappy ones depends on how involved they are in the neighborhood. I asked if they were active in a list of possible organizations: their neighborhood association; local government; a political party; any artistic or cultural organization; any environmental organization; any social justice organization; any athletic organization; or any religious organization. Consistently, the group of people who greatly liked the neighborhood were more active in some local organization. In each category, a significant minority of the happy residents took part, while almost none of the unhappy residents did. The happy residents were also more likely to say that they knew all of their immediate neighbors. The people who liked their neighborhood a great deal were much more likely to report that they were "very happy," whereas very few of those who disliked their area were very happy. This fits with the general findings of happiness research: people who work with others on a project larger than themselves are more likely to be very happy (Weston, 2015).

Each of these neighborhoods has become famous, a kind of project in its own right. Residents have their insiders' view of what the neighborhood is like; they are also aware of how their neighborhood is viewed by outsiders, especially by suburbanites. I asked each respondent to select from the following list which terms *you*, the resident, think apply to your neighborhood, and then to choose which terms *people from the suburbs* would use to describe the neighborhood under study. The terms were:

Vibrant __ Yuppie __ Hippie __ Friendly __ Nosy __ Interesting __ Creative __ Edgy __ Cool __ Hip __ Trendy __ Hipster __ Bohemian __ Bourgeois-bohemian (bobo) __ Weird __ Snobby __ Diverse __ Tolerant __ Crowded __ Kid-friendly __ Convenient __ Expensive __ Affordable __ Sketchy __ Dirty __ Dangerous __ Up-and-Coming __ Over __

Table 11.4 shows the three terms the residents who most liked their neighborhood chose most often to describe where they lived. The first column shows how the residents themselves feel about their neighborhood. The second column shows what those happy residents think suburbanites view those same neighborhoods.

A main reason many suburbanites give for not wanting to live in "the city" is a fear of crime. The people who like their boburb, on the other hand, are not very worried about crime in their neighborhood. While every place has some risk, they know that every place is safe if you know the rules. The survey asked how much they agreed with the statement "I worry about being a victim of crime in my neighborhood." Among the happy residents, half were not worried about crime, while only a third reported being worried at all. When they listed what they believed suburbanites thought about these boburban neighborhoods, only a few checked "dangerous." Wicker Park is the exception, with a third worried about crime, only a quarter not worried, and a fifth believing that suburbanites looked on Wicker Park as dangerous.

What comes through most strongly about the appeal of all of these boburbs is that they are *vibrant, creative, cool, interesting,* and, especially, *friendly.* This last quality is most important to the residents. Though they are urban neighborhoods, where people are "on top of one another" much more than in the suburbs, the people who like the boburbs revel in that greater social

Table 11.4 Top descriptors of various boburbs

Neighborhood	What the happy residents think about the neighborhood	What they believe suburbanites think about the neighborhood
Travis Heights	Friendly, Creative, Expensive	Expensive, Convenient, Cool
Wicker Park	Vibrant, Friendly, Convenient	Cool, Friendly, Diverse
Fishtown	Creative, Vibrant, Friendly	Up-and-Coming, Hipster, Diverse
Virginia Highland	Friendly, Creative, Convenient	Expensive, Friendly, Interesting
Pearl District	Friendly, Creative, Expensive	Expensive, Trendy, Interesting

density. They like interacting with one another. While only 30–40% grew up in a physically dense neighborhood, 80% preferred it now. Clearly, the people who choose the boburb choose it not just for convenience or material benefits, but for the kind of social life the boburb makes possible.

References

Bill Bishop, 2008. *The Big Sort: Why the Clustering of Like-Minded America is Tearing Us Apart.* Boston: Houghton Mifflin.

Morgan Brennan, 2012. "America's Hippest Hipster Neighborhoods." *Forbes. com*, September 20. https://www.forbes.com/sites/morganbrennan/2012/09/20/americas-hippest-hipster-neighborhoods/#21c49922cb38.

City of Atlanta, Department of Planning, Development, and Neighborhood Conservation; Bureau of Planning, 1988. "History of Virginia-Highland." Compiled June. Retrieved 2018 from https://web.archive.org/web/20040602180005/http://www.vahi.org/pdfs/history.pdf.

Markman Ellis, 2004. *The Coffee-House: A Cultural History.* London: Weidenfeld and Nicolson.

Jeff Huebner, 1997. "The Last Drop: Urbus Orbis, the Wicker Park Coffeehouse and Cultural Landmark, Falls Prey to the Gentrification It Helped Attract." *Chicago Reader*, November 20.

Bruce Johnson, undated; retrieved 2018. "Visions of the Pearl." Blog. http://www.visionsofthepearl.com/the-cultured-pearl/45yt8ukby21g87nmhysjoecju6ke6w.

Emily Kleine, 2001. "Virginia-Highland: Classic Homes and Convivial Atmosphere Reel 'Em In." *Creative Loafing*, January 27.

Richard Lloyd, 2006. *Neo-Bohemia: Art and Commerce in the Post-Industrial City.* New York: Taylor & Francis.

Joshua Long, 2010. *Weird City: Sense of Place and Creative Resistance in Austin, Texas.* Austin: University of Texas Press.

Geoffrey Moss, Rachel Wildfeuer, and Keith McIntosh, 2017. "Bohemian But Not Anti-Bourgeois: An Artistic Bohemian Lifestyle Community in Philadelphia." Presented to the American Sociological Association, Chicago.

Charles Murray, 2012. *Coming Apart: The State of White America, 1960–2010.* New York: Crown Forum.

Niche.com, 2018. Retrieved October.

Peter Lane Taylor, 2018. "How Fishtown, Philadelphia Became America's Hottest New Neighborhood." *Forbes*, May 2.

William Weston, 2015. "Happy, Busy Calvinists." *Society*, 52, 6 (May/June): 378–382.

Wikiwand, 2018. "Keep Portland Weird." http://www.wikiwand.com/en/Keep_Portland_Weird

Part IV
Between bohemia and suburbia

12 A measured appreciation of the boburb

The chronic irritant of boburban life is parking.
The chronic irritant of suburban life is barking.

The overwhelming virtue of the boburb is that it is more interesting to adults and youth. Most people with a choice like to go there sometimes, and many people I talked to wanted to live there when they were not raising small children, both before and, at least in imagination, after. And if they were never going to raise small children, the boburb was their first choice all along.

The great and overwhelming virtue of the suburbs is that they are the safest place to raise small children. This is why most families with a choice overwhelmingly vote with their feet, generation after generation, to move to the suburbs (Kotkin, 2016). Raising the next generation is the single most important project of any society. The suburbs are where most middle-class people live in the small-children phase of life.

Whether you think it would be easier or harder to raise teenagers in the suburb or the boburb depends on how fearful you are in general. For the more fearful, teenagers are more likely to seek dangers than little kids are, so it is more important to keep them in a controlled environment. For the less fearful, teenagers can do more things for themselves, so it is less important to control their environment, and a positive good to encourage them to develop independence.

The suburbs are better at their one job than the boburbs are at any one of their many jobs – except the meta-project of fostering a diversity of projects. The suburbs are a more-sought choice for those families engaged in the project of raising small children. By the same token, though, families are less loyal to their suburb when that project is done. The boburbs foster intense place-attachment because they foster a variety of projects throughout the life-course.

The suburbs are a monoculture; the boburbs are the city's most diverse ecology.

The suburb is built on an intricate system of controls. The boburb fosters a rich mix of connections.

Every kind of place has the vices of its virtues. Let us first consider the virtues, and related vices, of the suburbs.

Suburbia as the place to raise children

In America today, suburbs are thought of as the best place to raise children if you can afford it. This is true from near the top of the class structure, where the working rich live in McMansion gated communities, to the poorest inner-city families, who dream of moving to white-picket-fence suburbia, and all the classes in between. This dream is shared by many English-speaking countries, following a British model. On the continent of Europe, and in places shaped by that model, by contrast, the bourgeoisie live in the center of the city and push the poor to the suburbs. In the continental model, neither place is especially child-oriented (Fishman, 1987).

The suburbs also serve a number of other functions besides childrearing. They are a way of living partly in nature. They are a way of making exclusive communities – exclusive by race, religion, ethnicity, and, most effectively, by class. They are a way of putting property ownership in the reach of the masses – and following from that, a way of preserving the value of that property by controlling all other uses and residents of the neighborhood. They are a way of getting away from the congestion, noise, and dirt of the city – and from the isolation, boredom, and dust of the country. We normally think of the suburbs as primarily residential, but they have become the prime space for retail life, and increasingly for manufacturing and service jobs. Stores move to where the people are. Jobs move to where square feet are cheap.

The Anglo-American suburb brings together the ideal of an independent family devoted to raising their children with the material security of owning real property. The single family in a detached house with its own yard has become so fused with the middle-class ideal that there is no second choice (Perin, 1988, 30). As a result, the great bulk of new housing built by Americans for a *century* has been wave upon wave of detached suburban housing, spreading out from the core cities deeper into the countryside. These edgeless suburbs flow into the edgeless suburbs of other cities. They draw jobs and stores with them, dispersing the earlier functions of core cities. The core cities are hollowing down to a few government and economic functions, and residential neighborhoods of poor people who wish they could move to the suburbs. The great counter-trend to the movement of suburbanization is the boburb.

Historically, the American suburb has been defined by the norm of white, heterosexual, married couples raising their children. People who are related to that standard support the norm. People who do not fit that norm live somewhere else – especially in bohemia. However, whiteness, heterosexuality, and even marriage are not *essential* to the suburban norm. As non-white people who otherwise fit the norm were finally allowed to move to the suburbs, racial barriers began to break down. As homosexual couples who otherwise meet the norm move to the suburbs, the same happens to sexual orientation barriers. Stably cohabiting couples raising children are now widespread in suburbia.

Childrearing is really the most powerful tool of assimilation. Non-parents who help with the neighborhood kids are welcome. Single parents can be assimilated more readily than anti-kid adults. Single people looking to be married are readily incorporated, and the neighbors will try to help them reach that "normal" estate.

On the other hand, categories of people who are *not trying* to approximate the parental norm can't readily be assimilated in a suburb. A single person settled in singleness, who is opposed to marriage, of whatever sexuality, will have a cold welcome. They are a weirdness to be noted, watched, and avoided. And adults who are thought to be a danger to children are the worst impurity in the suburb, the most dangerous to what is held sacred.

Safety is for children

Safety is often the first reason people give for why they moved to the suburbs, most especially if they have children. The design of suburbs contributes to this sense of safety. The winding roads and cul-de-sacs are designed to slow down cars. The few entrances and exits are designed to keep out through traffic. The back yards, often fenced, are designed to keep children out of the streets and under parental eyes. The lack of any public places, often lamented by suburban critics and suburban children alike, is designed to remove any reason for outsiders to be there. As a result, outsiders are easy to spot and watch. If they don't have a good reason to be there – visiting a resident family, or providing a service to residents – outsiders are likely to draw a quick call to the police. From a suburban resident's perspective, policing outsiders is what the police are for.

Suburbs used to be very white. Non-white people, therefore, were easily spotted as outsiders, and in many cases all non-white people were presumed to be outsiders. Skin color was used as a handy and highly visible marker of safe/not-safe. Today, the suburbs are increasingly diverse by race and ethnicity – more so than popular perception has grasped. In the United States as a whole, most Asian Americans are suburban, and more than a third of Hispanics and African Americans are, as well. Some subdivisions are predominantly non-white. More are still predominantly white, but with a substantial minorities of various "minorities." Immigrants are increasingly likely to head straight to the suburbs, bypassing the historic pattern of ethnic enclaves in the city as the first stop (Teaford, 2008, 58ff).

Non-white suburbanites move there for the same reasons white people do – first, as a good place to raise their children, second for the economic security and status achievement that a house in the suburbs tends to mean. If anything, non-white suburbanites are likely to be *more* bourgeois in their habits than their neighbors are, because they are not protected by the privilege of other people assuming that they belong in the suburbs.

The assimilation of gay and lesbian households into the suburbs is even easier, especially if they are married and have children. The normalization of homosexuality and same-sex marriage means it is much harder to assume that "gay" means "dangerous for children" – the bottom-line concern for suburban culture. And same-sex couples who are raising children themselves have much more in common with their suburban neighbors than the few things that make them different.

The exclusion that is inherently built into suburban life is *class exclusion*. The threat that lower classes pose is both cultural and economic. The cultural threat, which is partly real and partly imagined, is that poor people are more likely to live in a non-bourgeois way. The suburbs fear people who are not quiet, orderly, self-sufficient, and non-confrontational. However, actual suburbs have people who are financially strained – single mothers after a divorce, for example – but who maintain the local standards of decorum. The image of poor people as bearers of "ghetto culture" or "redneck culture" are based on exaggerated fears fostered by the very social isolation that suburbs create.

The economic foundation of suburban life, though, is a bargain among all the residents to keep up the neighborhood's property values (this sections draws from Weston, 2018). Suburban subdivisions in America are, overwhelmingly, "deed restricted" neighborhoods. This means that all the properties come with covenants, codes, and restrictions, which strongly limit what property owners can do. The subdivisions are managed by a homeowners association (HOA), which has the power to fine and lien individual properties. The explicit aim of these rules is to keep up the property value of the whole neighborhood (Langdon, 1994, 87ff).

This is the structural obstacle to poor people living the suburban life – it costs a middle-class income to live there *by design*. And the richer the suburb, the more this is true and the more seriously the residents defend their property values. Suburban governments have been ingenious in resisting inexpensive housing, whether subsidized or not. Indeed, many suburbs legally incorporate themselves as little cities in order to resist urban regulations that might undermine their property values.

Protecting property values is also part of raising children well. For middle-class people, their house is usually their largest investment and greatest asset. While suburban parents do not usually expect to pass their house on to their children, they do expect to pass on the value of their house. We would expect suburban middle-class families to be the fiercest defenders of property rights because they are at the intersection of having a little property and having a posterity to protect.

The suburbs are where married people go

Bourgeois children may have a mildly "wild" period in their young adulthood. This is when they are most likely to live in the boburbs, go to bars, party around a work schedule, and look to settle down with a mate. They often look back on this period of life with fondness, but also as a cautionary tale about the dangers that could easily befall the vulnerable – especially their own children.

The transformation of these mildly wild youth into responsible adults is often experienced by them as becoming aware of the world's dangers and of the fragility of life. And the most powerful agent of this transformation is getting married and having kids. These are two separate changes. Marriage – living for another – can be enough to get people to see a good reason to be more careful and less reckless. And here "people" means primarily "men." The effect of parenthood is even more pronounced, especially for mothers. Put the two together, and married parents are the kind of people most devoted to the project of making a safer world, or at least of finding a safe corner of the world.

The project of the suburbs is to make a safe world for kids. The people engaged in that project are primarily married parents in the suburbs. They can see all the ways suburban culture affords to reduce risk, and thus to reduce their own anxiety. Their children do feel protected, but sometimes also stifled and bored.

A century ago, or even half a century ago, the suburban project was favored by patriarchal "family men." The virtue of a man taking his family to a separated house with a little land, away from the dangers of the city, is to protect them. The separated house seems like a more defensible castle. The little bit of land may even seem like a potential food production site. And the separated house in the separated neighborhood became the most reliable source of wealth accumulation for the family, because the very virtues of the suburbs for one middle-class generation were likely to be valued by the next (Marsh, 1990).

The separated house comes in a compound of separated houses. The compound is designed to make it easy to notice strangers, and to make other dangers less likely to come in.

The main act of protection that middle-class men do for their families is to make enough money to move them to a safe neighborhood. Neighborhoods in which the men watch out for one another's families and property are the safest. The pioneers in a suburb often do that. But even if the neighbors are not close, do not know one another, the very structure of the place is still fairly effective passive security. And neighbors who are not close might nonetheless be suspicious enough to notice "people out of place."

The first flaw in the suburban security plan is that it was a net loss for housewives. True, they gained from a feeling of security – if anything, mothers thought of more things to fear than their husbands did. But they lost easy access to everything else. And that was before most of them got jobs outside the house. Once most mothers worked, the very isolation of the suburbs – which was meant as a feature, not a bug – became a further obstacle to women having it all. Managing kid minding, which would be easier to arrange in a denser, stroller-able neighborhood, became a major task for working mothers, which took much of their income, free time, and mental energy.

The second flaw in the suburban security plan is that protecting kids by removing them from all sources of harm also means removing them from all sources of interest and agency. Their mothers could, at least, solve their own isolation by driving to what they wanted to do. Kids, on the other hand, were trapped in walking, or at most, biking range.

And then the divorce wave swept through the middle class. The above problems multiplied.

That is the experience of the suburbs for the best off. The other side of the looking-glass meant that middle-class non-white people could not reap the benefits of the suburbs. They either stayed in the more dangerous neighborhoods, or moved to suburbs that were less favored – by roads, public transportation, housing values, etc. This has had an important wealth accumulation effect, which carries over for generations. Also, non-white women worked in larger proportions than white women to begin with, which either kept the families from moving to the inconvenient suburbs in the first place, or faced non-white families with the "having it all problems" sooner and greater, and usually with less income to cover it. The divorce wave hit them even harder. And their kids looked like the very thing most suburbanites feared, so they did not reap the same protective benefit.

A further problem with the suburbs is that it multiplies the fears that it was designed to avoid in the first place. Not having contact with decent non-white people makes suburban white kids more fearful of all non-white people. Not reaping the benefit of the city makes suburbanites fearful of "the city" in general. *The physical isolation, which was designed as a protective feature, becomes a cause of mental isolation, which is a costly bug.*

Understanding suburban culture

My initial plan was to interview people I knew in various kinds of neighborhoods, then have them introduce me to their neighbors. This worked very well in the boburbs, and led to many fruitful discussions. In the suburbs, though, I ran into a barrier again and again: it would violate suburban norms of privacy to impose a stranger on your neighbors. To be sure, everyone I interviewed had civil relations with most of their neighbors, and quite a few had friends in the neighborhood, especially through their children. A few people I interviewed were willing to connect me with their friends who lived in other suburbs. But in only one case was I able to go from one suburbanite, through a mutual invitation, to an interview with a neighbor. Even in the few cases when a mutual introduction by email was made and the second family said yes, it turned out to be a "Southern yes" – which was a polite way of saying no by never actually setting a date to meet.

The suburbs draw people who want to do a good thing for their families, and *not* have to think about it all the time. The suburbs have become the default option for raising kids, and like all default options, it has an elective affinity for the incurious. The suburbs have become the "best option" for raising kids, so people who think about difference only in order to rank them don't need to think about where they live any more, once they have picked the best option. They do not have to weigh the pros and cons of every kind of place, nor see that the virtues of any choice come with attendant vices. Indeed, they don't want to have to think about it anymore, which is why they are suspicious of people who ask them to justify their choice.

The suburbs were not meant to be a distinctive kind of community. They were built piecemeal as housing, which was chased by retail, which in turn drew other jobs, which is now trying to be a culture generator. For most people who live there the suburbs are a compromise. For a few they are the best of both worlds between country and city. For most they are a second best for people who would prefer one or the other. A subset of suburbanites want a small town, and imagine that they could have that in a suburb. The people who speak most fondly of suburbs are people who grew up in them, talking about quiet cul-de-sacs where a group of kids could play.

Some people want a new, managed community in which "they don't have to do anything." In my interviews these were primarily men, the kind who appreciate the freedom from choice that uniforms bring (Fussell, 2002). What I call the "corporate perfect" aesthetic, especially when maintained by an outside maintenance company, is appealing not primarily because it is *beautiful*, but because it is *socially correct*. It has an assured status, and does not require thought or drama with others about your choices.

The fertility advantage of the suburbs is obvious (Egan, 2005). In my survey of Centre College alumni, half of those living in the boburb had no children, compared to only a quarter of those in the suburbs. The boburbanites are, remember, younger than the suburbanites, and no doubt many of those living childfree now will have kids later. The point, though, is that the people who do have children move to the suburbs.

The suburbanites are also more likely to have larger families. Comparing just parents with parents, the same proportion of families in suburb and boburb have only one or two children. When we get to families with three or more children, however, the suburban portion is double the boburban (Table 12.1).

The suburbs appear to be more fertile than the boburbs. This is hard to tell from survey data or even from Census data, which only asks how many children are in the household now, not how many the residents have ever had. Still, it makes sense that people who have more children are more likely to choose the suburbs to raise them in. You can get more house for the money, and a more child-oriented neighborhood.

Nurturing more children is perhaps the greatest gift of the suburbs to society as a whole. It is no mean contribution at a time when all advanced societies are realizing that the problem of the future is not going to be too many people, but too few (Wattenberg, 2004; Hvistendahl, 2011; Last, 2014).

Table 12.1 Number of children by neighborhood type

Kids	None	1	2	3	4+
Live suburb	26	15	28	23	8
Live boburb	46	13	25	12	4

The boburban project is diversity

The boburb, at its best, is a kind of neighborhood that preserves the vitality, diversity, and liberty of bohemia, while offering some of the security and durability of suburbia. Boburbs are not inevitable, and not every city ever has one. The most likely path to a boburb is for it to emerge from a bohemia that works, that draws enough bourgeois bohemians to stabilize the creative energy of bohemia. Bohemias are inherently ephemeral. A chronicler of bohemia wrote "In America in the 'Nineties, unalloyed garrets were rare and short-lived. Rich folks moved in as rapidly as these happy areas were dis-covered by the obliging rumor of the press and the novel" (Parry, 1933, 183). What is important to note here is that he is talking about Chicago in the *1890s*. Bohemia is always already over. The danger for a boburb is that it will succeed too well, will become so stable, secure, and routine that it is like every other suburb.

Boburbs have economic effects and political leanings. They are sought after as incubators of the new economy. They are the home culture of active pro-gressives. For both reasons, they are increasingly driven by cultural produc-tion. This is why they emerge from bohemias stabilizing, rather than from suburbs vitalizing. But the main aim of a boburb, or any residential neigh-borhood, is not primarily economic or political, but rather the daily stuff of domestic existence. Here the way in which boburbia is the middle position between adult-oriented bohemia and child-oriented suburbia is clearest.

A boburb is unlike most other residential neighborhoods because it delib-erately mixes uses and functions. Sites of economic production and economic consumption are within walking distance of the residences *by design*. This is what people who desire boburbs mean by "walkability", and why they cite "convenience" as one of the best features of their neighborhood.

The great virtue of the boburb is that it has the richest *mix* of social assets in the whole metropolis. It has places to live for a mix of classes, retail offer-ings for all kinds of goods, professional services, white-collar work, even some manufacturing. The boburb is often the center of education and culture pro-duction for the whole region. It is, of course, possible to make a good life in any kind of neighborhood. None has a monopoly on good things. And not everyone would want to live in the boburb – the city has a whole ecology of neighborhoods because different folks want different strokes.

The boburbs tend to draw people who create what we might think of as the meta-asset of a city: the analysis of the whole city itself. These are the people who consider the ancient problem of the *polis* – as you are doing right now. Aristotle famously said that we are *zoon politikon*, which we usually translate as "social animals." More literally, we can read that as "creatures of the polis." The polis, in ancient and in modern thought, includes the whole political economy of the region that the city leads. So, though modern cities are much larger than ancient ones, and the suburbs and exurbs extend much farther out into the agricultural hinterland and the forests beyond, the whole long conversation about flourishing in the polis applies as much today as it did for the ancients.

Who thinks about Louisville as a whole? We can start with the Center for Neighborhoods. Located downtown, the Center is a non-profit that studies and promotes flourishing in all the neighborhoods in Louisville. The staff focus much of their attention on building up the poorer and less organized neighborhoods. They try to have full connections with the suburbs, but few suburbanites see sustaining the whole region as their problem. The Center for Neighborhoods has its strongest contacts in the Highlands and Crescent Hill. The neighborhood associations there are the most vibrant. Moreover, those neighborhood associations supply the partners for the Center in building up other neighborhoods. When the Center hosts its Neighborhood Summit for all the neighborhood leaders in the city, the only neighborhood organizations to co-sponsor the event are Cherokee Triangle and Crescent Hill. The staff of the Center predominantly live in the boburbs or bohemian neighborhoods; they are much less connected to the suburbs. The founder of the Center lives in Cherokee Triangle.

The do-gooder non-profits tend to be based in the poor places they serve. Their staff, though, live in the Highlands and Crescent Hill. The younger ones live in Germantown and Clifton, especially if they do not (yet) have children. The civil servants most involved in overcoming racial disparities in the city live in the bohemianizing inner neighborhoods of Limerick, Merriwether, St. Joseph, Old Louisville, Schnitzelberg. If they ever lived in the suburbs, it was by accident. They move to the "real" city to be part of the action they are remaking. They do not live in the poorest neighborhoods – even if they were raised there. But they choose to live on the vibrant edge of the revitalizing polis.

The activist churches are much more likely to be in the boburbs. Many of their members live in the suburbs, but come in to the city to take part in a church that defines itself by service to the city's neediest. Often those members became associated with the church when they were emerging adults, living in the boburb, who then moved out to the suburbs when they had kids. But they brag about how many churches of their denomination they pass on the way in from the suburbs, to emphasize their commitment to the service mission of the church. Many of the boburban churches are notably liberal theologically, especially on queer issues. Highland Presbyterian Church started a ministry, now called the Presbyterian Community Center, to "Smoketown's African-American residents" in the 1890s. In the great struggle of the

20th century for desegregation and civil rights, Highland Presbyterian came out early for desegregation and decided to stay in the city even though most members lived in the suburbs. Two successive mayors of Louisville in mid-century were members of the church, as was the judge who later wrote the city's open housing ordinance. Today, Highland Presbyterian hosts Kentucky Refugee Ministries (Raymond and Ellison, 2008, 80ff).

Theologically conservative churches in the boburbs tend to either become more liberal, or depart to the suburbs. Highland Baptist Church was begun as a preaching station for Southern Baptist Theological Seminary. It was thought of as "aristocratic" among Southern Baptists – the seminary presidents lived down the street, and many seminary professors were members. It tended to take the moderate side in the denomination's religious conflicts. In the 1980s, Highland Baptist switched to the more liberal Cooperative Baptist Fellowship (Smith, 2005). Sojourn Church bought a large Catholic building in Shelby Park, on the border between Germantown and the black ghetto. This large, active church is a theologically conservative, but socially liberal Southern Baptist ministry, but they soft-peddle their connection to the Southern Baptist denomination. Sojourn has been an agent of boburbanization, if not gentrification, in that frontier neighborhood. In their immediate shadow a new bakery has opened, along with that symbol of the boburb, a coffee roaster.

Youth as the hinge: are the boburbs the best place for teenagers, or the worst?

> The slogan of the boburb is "Keep Louisville Weird."
> The slogan of the suburb is "Keep the Swim Team Full."

The swimming pool, and the swim team, is one of the few reliable solidarity building institutions in the average American subdivision. This is especially true for bringing teenagers together. Running the pool is the most time-consuming task for many a HOA. Recruiting the team is a constant project, because the team members age out so quickly. Usually they then go to college or start adult life – somewhere else. A neighborhood swim team is a consuming affair for swim parents as well as the young swimmers. Unlike many other kinds of teams, swim practices are not limited to the school day, nor is the season limited to the school year. Swimming is overwhelmingly a middle-class and upper-middle-class sport, rooted in the suburbs. And it is an ingenious form of social control for youth – parents know their children will be under adult supervision before school and after, occupied on weekends, doing something good for their health, and in a sport which leaves them too tired to get into much trouble.

To be sure, some boburbs have swim clubs, too. The most famous swim club in Louisville, Lakeside, is located in an old quarry in the outer Highlands. Its competitive swim team, however, trains at their center in Crescent

Hill, or the exurban facility in Shelbyville, beyond the Jefferson County line. Norton Commons has two pools, but they are designed for family splashing, not competitive racing. The new YMCA at Norton Commons offers swim lessons, but not a competitive swim team. Instead, what it offers for youth sports is an inclusive Adaptive Sports Program in soccer and basketball

> designed so that any child and teen ages 5–14 with a physical or mental disability has the opportunity to make new friends, have fun playing sports and learn teamwork. Our rules are made to enable each child towards success.
>
> (YMCA Louisville, 2018)

Their theme is diversity and inclusion – very boburban values.

Earlier we met a large family, active in a suburban megachurch, who chose between Lake Forest and the Highlands. In many ways they are natural suburbanites. Yet they ended up in the Highlands because they wanted their teenagers to have "sidewalks that go somewhere." Again and again the boburb parents wanted their children to learn independence and cosmopolitanism. One mother was delighted that her son and his magnet school friends, who came from all over the city, would regularly take various buses to meet at a sports bar to watch the British football (soccer) team Arsenal. By contrast, when I asked a couple of Lake Forest parents if they would ever like to live in the Highlands, they said "not with kids" because they wanted to control the social values of the people their kids hung out with. When I asked if, even in their gated community and their children's religious school, they vetted their kids' friends, they vigorously agreed.

This contrast about which is the best way to raise teenagers gives us some insight into the "Louisville question" that is so puzzling to transplants – "Where did you go to high school?" The city's rich spectrum of public schools and array of private schools means that each high school symbolizes a class fraction, and suggests an ideological bent. In the past, which high school one attended would also give a clear signal about race; this is less true today, especially among middle-class families of all races and ethnicities. The knowledge class in the city relies on the public magnet schools, especially Manual, Male, and Atherton. The corporate class in the suburbs finds the east end public schools, Eastern and Ballard, to be acceptable. In the private system the strong Catholic options are mostly in the boburbs – St. Xavier and Trinity, Sacred Heart and Assumption – as is the oldest secular private high school, Collegiate. The high schools with the most conservative reputations are in the far eastern suburbs, namely Kentucky Country Day and Christian Academy of Louisville. And the two most liberal high schools – St. Francis Episcopal High School, on the private side, and James Graham Brown School, on the public side – are also the most urban of the college-oriented high schools.

A woman with a unique perspective on these schools gives some insight on their shades of meaning. She herself graduated from one of the suburban public schools, then left town for college. When she returned, she wanted the vibrancy of the boburb. Later, when she had a child, they moved to a more suburban place. She taught in several private schools, including the most liberal and the most conservative, while her child went to a public magnet high school. When asked about the political differences between the suburban and urban school, she said the red/blue split was definitely true for the parents, less so for the teenagers. The urban school parents, many of whom lived in the suburbs, had more progressive beliefs, while the suburban school parents were "older money" which they were trying to preserve in the next generation. The urban school kids took more risks, and were less set on making money. The suburban school students wanted to live like their parents, which made them conservative in several respects. She routinely asks her suburban school students "Which matters more, happiness or money?" They are split down the middle in their responses. For her own part, after their children finished school, she and her husband were happy to move back into the city; their son lives in Germantown.

Spending privilege vs securing status

Boburbs appeal to those secure in their middle-class social status. People with greater "seniority in the bourgeoisie" (Bourdieu, 1984, 15ff) – multiple generations of economic security and higher education in this country – can afford to risk a little confusion about their status that might come from living in or near bohemia. They know their own lineage, and expect others to credit their status, as well.

People who are new to the middle class, or who fear that others might not credit their status, are more likely to want a big house in a middle-class neighborhood far from the city. Immigrants and first-generation college graduates who want to send an unambiguous signal of having "made it" – to themselves as well as to others – embrace the newest and richest suburbs. An immigrant couple with several children built a large house – more than 5,000 square feet, with six bathrooms – in an exclusive neighborhood in the far edge of the county. They drove expensive cars, and sent their children to private schools. Though both were professionals, they could not be certain that other people would accredit their achievements and status. They told me frankly, "What is the point of having money if you don't show it?" Another kind of immigrant, a first-generation college graduate from Appalachia, said he moved his family to the suburbs because that is where the malls were, and from his childhood perspective, that was the height of class. As his career advanced, he moved his family to Lake Forest.

By contrast, most of the people I talked to in the boburbs were multiple-generation collegians. They had no concern that their smaller house, or use of public transportation, would make others doubt their status. Several families who had moved to the Highlands from expensive suburban subdivisions felt that there was more concern with display of money out there – even though there was also considerable money in the boburbs. In recent decades there has been a "bright flight" to the most interesting parts of cities (Speck, 2012). Whereas in the 1960s, the suburbs were more educated than the cities, since the 1980s the balance has clearly shifted the other way (Beeson, 2012).

There is a strong relationship between education and income for our focus neighborhoods. Overall in Louisville Census tracts there is a 0.83 correlation between average education level (percent with a bachelor's degree or higher) and average income. Our most bohemian neighborhoods, Germantown and Clifton, had slightly above-average levels of education, but below-average levels of income, due, in part, to their admixture of working-class people and low-earning, but college-educated artists. Both the boburbs and the suburbs have above-average education and earnings, compared to the city average. However, the Highlands had the highest *education* level – 1.7 standard deviations above the city average; whereas Lake Forest had the highest *earnings*, more than three times the city average. Compared to the citywide averages, Crescent Hill is 1.1 standard deviations higher in education than in earnings, whereas Lake Forest is 1.6 standard deviations higher in earnings than education.[1] *The boburbs out-educated their earnings, while the suburbs out-earned their education.*

A nuance of who lives in the boburbs and who lives in the most expensive and exclusive suburbs turns on their "composition of capital." While both groups are middle class or upper-middle class, they differ in how much of their class status comes from money, and how much from culture. The monied fraction of this dominant class are likely to invest in exclusive goods – expensive houses and cars, country clubs, private schools. The cultured fraction will spend a bit less on their houses and cars, go to the coffeehouse for leisure, and invest in the magnet public schools (which are likely to be in their neighborhood to begin with). The experiences of cosmopolitan culture can be acquired with time spent in learning about them, even for people who don't have as much money to spend (Bourdieu, 1984). The boburbs appeal to the cultured fraction because there is more concentrated and original culture there.

The emphasis on culture, even more than politics, may be what sets bohemia and the boburbs apart from the suburbs. The knowledge class clearly favors the boburbs, while the corporate class favors the suburbs (Weston, 2011). In both places there are educated people and people with middle-class incomes. In both places there are people with upper-middle-class (and more) earning *potential*. But the people who value producing and consuming knowledge and the arts were more likely to be strongly attracted to the boburbs. All of the artists I talked to either lived in the Highlands or Crescent Hill, or wished they could afford to. The educators, at all levels, tended to favor the boburbs; for example, of 16 sociology professors working in the city, eight live in the Highlands, Germantown, or the adjacent Old Louisville, while only four live in the eastern suburbs. The people who worked for nonprofits, both the culture-oriented and the justice-oriented, were strongly drawn to the dense networks of the boburbs and bohemias.

Bobos and bohos alike cited the cultural features of their neighborhoods as a main draw. They named houses that were interesting, inside and out, as a reason to pick a neighborhood – where suburbanites would cite "low maintenance" and "high resale value" as their houses' most important features. Quite a few bobos had lived in the suburbs, usually for their small children, but moved to the boburbs because they felt as if they were in a cultural desert. They disdained chain stores and malls, but instead liked unusual and eccentric stores.

Boburbia's greatest strength: social density

Americans have long worried that modern society is diminishing our sense of community. Even when we look back to what seem like times of great community feeling, people in those days lamented the loss of community compared to even earlier ages. America was born modern. We have always been a very mobile society, and never had an era of stable villages where most people were related.

Small towns come closest to having that sense of community. Most people, when asked, would like to live in a small town – or at least a place that *feels* like a small town (Brower, 1996). To be sure, there is also a well-developed counter-narrative of the young person who feels trapped in a small town and yearns for a more cosmopolitan and sophisticated life in a city. But most people – especially most parents – yearn for a secure community in which to practice their family life.

The Pew Polarization Survey data lets us compare two different versions of "where would you like to live?" – the suburb/boburb preference, previously discussed, and a different question about the respondents' preference for living in a city, suburb, small town, or rural area (Table 12.2). This shows an interesting nuance. At the extremes, those who want the city want the boburb, and those who want the country prefer the suburb. The people who want the suburb, not surprisingly, favor the suburb, though only slightly. The surprise is that the people who want a small town prefer the boburb to the suburb. I believe that what they want in both cases – boburb and small town – is greater density of social interaction.

Cities can't offer that kind of community to all. Some "urban villages" with a shared ethnicity create something of a small-town feel in the city, but that

Table 12.2 Neighborhood preference compared to urban-to-rural preference

Preference	City	Suburb	Small town	Rural area
Suburb	24	50	45	71
Boburb	75	47	53	25
Missing/ no answer	1	3	2	4
Total	100%	100%	100%	100%

usually only lasts a generation (Gans, 1962). Many have turned to the suburbs as real-world "bourgeois utopias" where like-minded people, with a shared interest in children (and, implicitly, a shared social class), can create true communities – paid for by city jobs (Fishman, 1987). But the kind of community the suburbs create depends more on shared ideas of what *not* to do, than it does of doing things together.

Suburbs do, in fact, often create a low-conflict environment that would be the envy of the dangerous places of the world. But open conflict is rare in the suburbs not because neighbors never annoy one another, nor because they all feel close to one another. Rather, suburbanites have a "moral minimalist" norm, based on *not* getting involved in one another's lives. Instead, they share a value of respecting privacy. Neighbors are loathe to confront neighbors, and even more reticent about confronting strangers. Suburbanites rely on the police, the zoning board, animal control, and, especially, HOA rules to indirectly control one another and diffuse conflict (Baumgartner, 1988; Low, 2003).

Many of the suburbanites mentioned that one thing they valued most about their neighborhood was that they did not *have to* interact with the neighbors when they did not wish to. To be sure, suburbanites often had neighbors they valued and did things with, especially through the neighborhood swimming and golf clubs. But the difference is that suburbanites only had to deal with people they *chose* to interact with. Their "community" institutions have formal rules about who is invited in, how much contribution is required, and which methods to use for expelling deviants. This is why the country club is an apt symbol of suburban social interaction.

The boburbs create social density through a web of weak ties, joining together long strings of overlapping acquaintance groups (Simmel, 1955). This ecology of weak ties is good for spreading information and fostering creative new groupings for specific purposes (Granovetter, 1973). Bohemia and boburbia are rich in "third places" – not home, not work, but gathering spots where strangers can hang out and become acquaintances. Parks, libraries, bookstores, and restaurants can serve this purpose. Even better, though, are bars, taverns, and, especially, coffeehouses, where people can linger and talk (Oldenberg, 1989). Of course, all of these things also exist in the car suburbs. But out there, you have to drive to get to them. Bardstown Road and Frankfort Avenue are lined with coffeehouses – where many of our interviews took place.

These nodes of social connection *overlap* in the walkable part of the city. The gathering place of one group is next door to the gathering place of another, and probably shares some members. All of the functions of life – including those involving children – are likely to be in the same walking zone, bringing together people from different stages of life (Speck, 2012; Montgomery, 2013).

This dense web of group affiliations is not, itself, designed by anyone, nor does it serve any one purpose. Instead, it is an organic emergent from a densely overlapping set of social interactions (Smith, 2015). What it is good for is spreading information, and generating voluntary associations. Alexis de Tocqueville's analysis of what makes American democracy work rests on the providential way in which a mass of self-interested people could come to see that their self-interest, rightly understood, is best served by banding together to address the problems of their society (Tocqueville, 2004). These voluntary associations are the wonder of American social life, and nowhere are they more abundant than in the boburbs. In the car suburbs you might drive in one direction to meet one friend, in another direction to interact with acquaintances who share an interest, and in yet another direction for work. In the walkable boburbs, these circles are more likely to overlap, and the people who are connected to you are more likely to be connected to one another, as well.

Social density is not only a boon to voluntary associations – it may also be a key component in economic development. Suburban growth has meant that current jobs are diffused out into the suburbs – to the point at which most people in a metro area work in the suburbs, as well as living there (Teaford, 2008). Some think this means that concentrating economic development in the core city, or anywhere, is no longer necessary.

A powerful counter-argument says that the "creative class" is building the future economy. The creative class gets its creativity from rubbing shoulders with other creative people. Moreover, the creative class likes to live among people who are creative in different fields than their own, and enjoy the fruits of their creative labors. Sociologists have studied the ways in which "scenescapes" of distinctive cultural forms define neighborhoods. The most vital scenes draw people, especially young, educated, creative people with money and/or great economic potential. Developers and city planners watch closely for the emergence of any cutting-edge scene, and jump on the chance to invest and develop it (Silver and Clark, 2016). The creative class needs social density to be creative. And the whole national economy of the future depends on the work of the creative class today (Florida, 2002). This was clearly true for the artists and musicians I interviewed in Louisville – frequent interaction with the "creative community" was vital to their work, and their happiness.

Boburbs and human flourishing

Human beings make their own neighborhoods, but not under conditions of their own choosing. The educated middle class have quite a bit of choice about what kind of neighborhood they want to live in and raise their families in. For more than a century, the choice of the American college-going class has been the suburbs. It is a premise of Tocquevillian democracy that self-interest, rightly understood, will lead most people to choose the option most conducive to their flourishing.

To be sure, the vast expansion of the car suburbs has also been for other reasons than human flourishing, either of families or of society as a whole. In the overwhelming majority of cases, private corporations looking to make a profit built the suburbs. Indeed, fortunes have been made in land speculation for suburban development. Other economic interests benefit from suburban sprawl – car companies, oil companies, road construction firms, mortgage banks, and the many industries that serve a car-based population. Let McDonalds stand in for dozens of other firms that depend on car suburbs for their livelihood (Teaford, 2008).

Moreover, those firms, especially the "car lobby", actively worked to get government to subsidize the automobile suburbs and undermine urban redevelopment. The government supported new home construction with loans through the Federal Housing Authority and the GI Bill, but did not do the same for urban infill and rebuilding. The tax code was changed to allow companies to quickly depreciate new building, encouraging an endless cycle of even more new building further from the city. Home mortgages are deductible from federal income tax, but not rent. The government has a special tax for highways, but not railroads, and allowed the car companies to buy up and close the streetcar lines (Hayden, 2003).

Worst of all, the government allowed the racial discrimination which had segregated the cities before World War II to be spread to the suburbs after the war, even after the Civil Rights Movement victories should have made such discrimination a thing of the past. As "white flight" – and black middle-class flight – made the cities poorer, the suburbs created governments which resisted sharing development with their mother city. Suburban-based politicians found success with "law and order" campaigns that were anti-city (and implicitly racial) (Lipsitz, 1995).

There has long been a political divide between city and country. With the growth of the suburbs to a majority of the population, and an even greater power in the electorate, the "blue vs. red" divide has become even more polarizing. There has been a "big sort" since the 1970s, whereby people are increasingly likely to move to neighborhoods where the people are ideologically similar. The blue places have gotten somewhat more liberal, and the red places have gotten much more conservative. And the blue/red divide is mostly a pro-city/anti-city divide. The liberal coalition unites urban poor people, highly educated cosmopolitans, and non-white people of all classes. The conservative coalition unites rural poor people, the economic elite, and security-oriented people of all classes (Bishop, 2008).

The highly educated cosmopolitans are concentrated in the boburbs. This is why they are the bluest spots in their states. Their political concerns are not confined to their community or even the city as a whole. As such, they wield a disproportionate influence in creating policy at all levels. Their policies tend to promote the very virtues of the boburb: cosmopolitanism; tolerance; public goods; education. As a way of life, they believe in walking, talking, learning, trying things. They see density as a good thing, a necessary foundation for vibrant and connected social life. They see diversity as a good thing precisely because it undermines fear of the Other, as well as promoting that rubbing-together of difference so conducive to creativity.

The blue ideal emphasizes open and connected public life, which it hones through a dense web of weak ties and voluntary associations. The red ideal emphasizes private life, which it protects with a loose social order of equally private-minded citizens. The most thorough study of urban versus suburban neighborhood preference found that the three biggest dividers were a preference for:

1 Activity vs. tranquility;
2 Sophisticated vs. "down-to-earth" neighbors; and, most importantly,
3 Diverse vs. similar neighbors (Brower, 1996, 98ff).

Controlled access vs. openness to experience

The suburban experience is all about controlled access. The kinds of controls form a spectrum from the macro rules, backed by law, through all kinds of private property controls, down to micro etiquettes of separation.

At the largest scale, suburbs achieve control through zoning and redlining. Zoning rules are what make a subdivision only residential. Minimum lot sizes are used to prevent density and keep up prices. When the Supreme Court made it illegal for the government to directly exclude families from housing on the basis of race in *Buchanan v. Warley* in 1917, the government and suburban developers did the next closest thing. The government made maps of credit risks by neighborhood, using race as a criterion. Private mortgage lenders then used these maps to justify refusing to lend to redlined neighborhoods that were bad risks, or to lend to non-white people who wanted to move to a better neighborhood, because they would thereby change the credit-worthiness of the neighborhood. The Fair Housing Act outlawed that practice in 1968, as well as outlawing exclusions against poor people. Since then, suburban communities have developed ingenious ways to prevent housing that poor people can afford, without specifically prohibiting the poor people themselves.

When deed-restricted communities were created, the government handed to the subdivision the tools of controlled access. This made the process of exclusion more local, more democratic, and to at least look like it was more by each family's choice. No one was obliged to choose a deed-restricted community. Today HOAs are not permitted to exclude on the basis of race or the gender of the homeowner; in many communities the sexual orientation of the homeowner is also not a legally permitted basis of exclusion. The core concern, class exclusion through property values, is the heart of what suburban exclusion is about today – and, I think, always has been. The fact that the great majority of housing in the suburbs is now deed-restricted is treated as an unintended consequence of many private choices, rather than the predictable outcome of class-based public policy.

Many other kinds of controls are common in the suburbs based on thousands of separate private-property choices. Country clubs are only the grandest of the many kinds of private, members-only facilities common in the suburbs. Suburbanites were more likely to assume they would use private schools, rather than be subject to the uncertainties of the public school assignment system. Some moved out of Jefferson County altogether in order to control which public school their child would attend.

Many houses have private security, and sometimes whole subdivisions will go in on sharing private patrols. The miniscule police forces of the small home-rule cities are, in effect, private security. There is almost no public transportation in the suburbs, which means all spaces and most social inter-actions are dominated by private automobiles. Many suburbanites have an active social life, but it is conducted mostly in private spaces, by invitation. Adult sports are conducted in private clubs. Even youth sports, once domi-nated by public schools, are being displaced by private leagues and travel teams. Colleges now recruit mostly from clubs, rather than high schools, in nearly every sport but football and, to a lesser extent, basketball (Luidens, 2005). Even the dogs, so important in creating social connections for dog walkers and other pedestrians in the boburbs, are more likely to live in pri-vate, fenced back yards in the suburbs.

At the most basic level of everyday life, suburbanites control who they interact with by distance, cost, and hypervigilance. Suburban places are farther from one another – too far to walk, and deliberately not connected to the forms of public transportation that poor people would use. One friendly HOA wanted to build a walking trail along a creek, but some of the homeowners on the trail feared that if there were a path, "undesirables" from outside the neighborhood would use it to come threaten their homes. A father in the far eastern reaches of the county allowed that his wife likes gentrifying, up-and-coming neighborhoods, like Butchertown. However, she will have to wait until the kids are gone, because in Butchertown, they would not feel the kids were safe to walk down the street. "If there's a lot of people coming by," he asserted, "then there will be bad people." His current far suburb was better because no strangers could come through unseen.

Suburban houses have been getting bigger and bigger with each generation, and "McMansions" are huge inside, which has the effect of making the whole neighborhood too expensive for any but the middle and upper-middle classes. There aren't just some big houses – there are only big houses. And beyond all these controls, the suburbs are the place where hypervigilance about any possible threat can become the norm, especially against perceived threats to children.

The desire for control in the suburbs gives a clue about what became a secondary research question in this study: why is it so easy to get interviews in the boburbs, and so hard in the suburbs? In any coffeehouse in the boburbs, strangers would be happy to tell me the history of their moves, tell me why they chose the boburb, and join in the spirit of the research project. In the suburbs, by contrast – well, in the first instance, there were no coffeehouses or similar places where I could meet strangers to ask the question. My suburban interviews were almost entirely with people I already knew, or through the personal recommendation of someone they already knew. Quite a few were the parents of my students, who did it as a favor to their children. After a while, I started to ask the suburbanites why it was so hard to get suburban interviews. Their answers follow the theme of controlling all access.

A conservative political professional thought perhaps that being interviewed feels like talking to the media, and "Republicans don't like the media." He also thought that where you live shows race and class, two subjects he and his colleagues did not like to discuss. A man with a working-class background who had worked up to a middle-class job and subdivision in the far east of the county offered that people moved away from others because they didn't want to get involved in general. About our interview, he admitted that he wondered "What is this guy after? What is he trying to sell me?" One Lake Forest couple even admitted that they were specifically suspicious of sociologists!

The single most telling statement made in any interview came from a young couple who were renting in Norton Commons before they married and bought a house. They had grown up in the suburbs. When they were just out of college, they lived in a suburb near Crescent Hill, where they and their friends would socialize in the bars and restaurants. They moved to Norton Commons when they were getting ready to settle into married life. They liked the community feel of the place, and the idea of a mixed-use neighborhood. However, when they looked at a condominium over a store, they realized they could see customers on the sidewalk through the windows. She said, "I just didn't like the idea of random people walking by my house." I reflected on this statement in every subsequent interview. The boburb residents all said some version of "I really appreciate the diversity of people who walk by my house." There, in a nutshell, is the contrast between an essentially suburban and an essentially boburban way of looking at social density and social difference. Are strangers a potential threat to be countered, or a potential resource to be interested in?

Why liberal boburbs vs. conservative suburbs?

Liberals live a more public life. Their ethical position is based on using government to care for the harmed, and create more equality (Haidt, 2012, 95ff). These two points are not identical, but are related. Quite a few boburbanites remarked that their life was like living on campus. They were content to have a smaller private space, close to others, because much of their life was lived in shared public spaces. In their book, this was a positive feature – interacting with other people is what makes life good. Public life is made possible by reliable social structures – the aggregation of individual choices is not enough to create a public life you can count on. And to be fair, these social structures need to be available to everyone. That requires government, because the market serves only those who can pay.

Conservatives live a more private life. Here the suburbs mix two different kinds of non-liberals, the social conservatives and the libertarians. They want somewhat different things, for which the subdivision is a reasonable compromise. The social conservatives want to preserve the traditional sacred goods of family life and nation. For a portion of American social conservatives, the "traditional" nation is a white nation. For libertarians, by contrast, the main and pretty much only sacred good is the liberty to do what you want and be left alone. They will put up with the potential intrusiveness of the HOA and their covenants, codes, and restrictions in exchange for the suburban etiquette of minding your own business.

The deepest difference: trust vs. fear

The question I began with was "why do people choose one kind of neighborhood or another?" As this research developed, the choice between the boburb and suburb became clearer and clearer. On the surface, they are equally viable options for the college-educated middle and upper-middle classes. Each has solid houses with durable value. Both are fairly safe, attractive, and are filled with "people like us." Yet there are differences that matter to where this class lives, and at what point in their lives. The boburbs trend younger and childless, the suburbs a little older and with children. The class fractions are different, with the knowledge class preferring the boburb, while the corporate class chooses the suburbs. At the next deeper cultural layer, the boburbs draw liberals and the suburbs draw conservatives, no matter what their stage of life or occupation.

The deepest layer of difference, though, is one of basic worldview.

If you trust the world, then a diverse neighborhood is itself an asset that accrues to you. If, on the other hand, strangers are dangers, then you must accumulate as many assets as possible to protect yourself. If you trust the world, then a tiny house in an interesting neighborhood is fine. If you fear the world, you need a big house to do all your living in. If you trust the world, strangers are interesting; if you fear the world, each encounter with a stranger incurs a transaction cost. Curiosity is cheaper than suspicion.

People who trust that the world is basically a safe place appreciate the diversity of the world, and want to experience more of it, right outside their door. People who fear that the world is full of dangers that might harm their family want to hold the world at a distance, control all interactions with it, and keep strange dangers far away. People who thrive in the boburb trust the diversity of the world. People who thrive in the suburbs take comfort in the security their space gives them from the dangers of the world.

Note

1 Based on subtracting standardized (z-score) earnings levels are from education levels. I am grateful to Nate Kratzer for this calculation.

References

M.P. Baumgartner, 1988. *The Moral Order of a Suburb*. New York: Oxford University Press.

Bonnie Beeson, 2012. "Suburban Advantage: Social Reality or Lingering Ideal?" MA Thesis, University of Washington.

Bill Bishop, 2008. *The Big Sort: Why the Clustering of Like-Minded America is Tearing Us Apart*. Boston: Houghton Mifflin.

Pierre Bourdieu, 1984. *Distinction: The Social Critique of the Judgement of Taste*. Translated by Richard Nice. Cambridge: Harvard University Press.

Sidney Brower, 1996. *Good Neighborhoods: A Study of In-Town and Suburban Residential Environments*. Westport, CT: Praeger.

Timothy Egan, 2005. "Vibrant Cities Find One Thing Missing: Children." *New York Times*, March 24.

Robert Fishman, 1987. *Bourgeois Utopias: The Rise and Fall of Suburbia*. New York: Basic Books.

Richard Florida, 2002. *The Rise of the Creative Class*. New York: Basic Books.

Paul Fussell, 2002. *Uniforms: Why We Are What We Wear*. New York: Houghton Mifflin.

Herbert Gans, 1962. *The Urban Villagers: Group and Class in the Life of Italian-Americans*. New York: Free Press.

Mark S. Granovetter, 1973. "The Strength of Weak Ties." *American Journal of Sociology*, 78, 6 (May): 1360–1380.

Jonathan Haidt, 2012. *The Righteous Mind: Why Good People Are Divided by Politics and Religion*. New York: Pantheon Press.

Dolores Hayden, 2003. *Building Suburbia: Green Fields and Urban Growth, 1820–2000*. New York: Vintage.

Mara Hvistendahl, 2011. *Unnatural Selection: Choosing Boys Over Girls, and the Consequences of a World Full of Men*. New York: Public Affairs.

Joel Kotkin, 2016. *The Human City: Urban Planning for the Rest of Us*. Chicago: B2 Books.

Philip Langdon, 1994. *A Better Place to Live: Reshaping the American Suburb*. Amherst: University of Massachusetts Press.

Jonathan Last, 2014. *What to Expect When No One's Expecting: America's Coming Demographic Disaster*. New York: Encounter.

George Lipsitz, 1995. "The Possessiveness of Investment in Whiteness." In Becky Nicolaides and Andrew Weise, eds. *The Suburban Reader*. New York: Routledge, 2006.

Setha Low, 2003. *Behind the Gates: Life, Security, and the Pursuit of Happiness in Fortress America*. New York: Routledge.

Donald Luidens, 2005. "NCAA Division III Athletic Recruiting and the Impact of Club Sports." *Liberal Arts Online*, 5, 11 (November).

Margaret Marsh, 1990. *Suburban Lives*. New Brunswick: Rutgers University Press.

Charles Montgomery, 2013. *Happy City: Transforming Our Lives Through Urban Design*. New York: Farrar, Straus, and Giroux.

Ray Oldenberg, 1989. *The Great Good Place: Cafés, Coffee Shops, Bookstores, Bars, Hair Salons and Other Hangouts at the Heart of a Community*. New York: Marlowe and Company, second edition 1999.

Albert Parry, 1933. *Garrets and Pretenders: A History of Bohemianism in America*. New York: Dover Edition, 1960.

Constance Perin, 1988. *Belonging in America: Reading Between the Lines*. Madison: University of Wisconsin Press.

Linda Raymond and Bill Ellison, 2008. *Like Jacob's Well: The Very Human History of Highland Presbyterian Church*. Published by the Church.

Daniel Aaron Silver and Terry Nichols Clark, 2016. *Scenescapes: How Qualities of Place Shape Social Life*. Chicago: University of Chicago Press.

Georg Simmel, 1955. *Conflict and the Web of Group Affiliations*. Translated by Kurt Wolff and Reinhard Bendix. New York: Free Press.

Christian Smith, 2015. *To Flourish or Destruct: A Personalist Theory of Human Goods, Motivations, Failure, and Evil*. Chicago: University of Chicago Press.

Peter Smith, 2005. *The Cloud of Witness: 110 Years of Faith – Highland Baptist Church, Louisville, Kentucky*. Copyright by the Church and the author.

Jeff Speck, 2012. *Walkable City: How Downtown Can Save America, One Step at a Time*. New York: North Point Press.

Jon Teaford, 2008. *The American Suburb: The Basics*. New York: Routledge.

Alexis de Tocqueville, 2004. *Democracy in America*. New York: Library of America. Translated by Arthur Goldhammer. Originally in two volumes, 1835 and 1840.

Ben Wattenberg, 2004. *Fewer: How The New Demography of Population Will Shape Our Future*. New York: Ivan Dee.

William Weston, 2011. "The College Class at Work and at Home." *Society*, 48, 3 (April): 236–241.

William Weston, 2018. "Are Neighbourhoods Real?" *Journal of Critical Realism*, February: 34–45.

YMCA Louisville, 2018. "Norton Commons." https://www.ymcalouisville.org/norton-commons/sports/special-needs-youth-sports.html.

Index